Degrees in Science and Mathematics: National Trends and State-by-State Data

Irene Harwarth
National Center for Education Statistics

U.S. Department of Education
Richard W. Riley
Secretary

Office of Educational Research and Improvement
Sharon P. Robinson
Assistant Secretary

National Center for Education Statistics
Emerson J. Elliott
Commissioner

National Center for Education Statistics

"The purpose of the Center shall be to collect, and analyze, and disseminate statistics and other data related to education in the United States and in other nations."—Section 406(b) of the General Education Provisions Act, as amended (20 U.S.C. 1221e–1).

November 1993

For sale by the U.S. Government Printing Office
Superintendent of Documents, Mail Stop: SSOP, Washington, DC 20402-9328
ISBN 0-16-042071-7

Highlights

- On the national level, bachelor's degrees in science and mathematics declined by close to 16 percent between 1985–86 and 1989–90, in contrast to an increase of 6 percent for all bachelor's degrees. During that same time, national numbers of bachelor's degrees dropped in the fields of engineering (14 percent) and computer sciences (34 percent).
- Bachelor's degrees in science fell in every state between 1985–86 and 1989–90. These decreases ranged from 2 percent in Wisconsin to 29 percent in Rhode Island.
- Bachelor's degrees in science also decreased as a percentage of all bachelor's degrees in every state between 1985–86 and 1989–90. In 1985–86, there were 279,455 bachelor's degrees awarded in science in the United States, over 28 percent of all bachelor's degrees. In 1989–90, 234,731 bachelor's degrees were awarded in science, 22 percent of all bachelor's degrees.
- In contrast to the decreases in bachelor's degrees in science, master's degrees in science increased in 37 states, and doctor's degrees in science increased in 44 states between 1985–86 and 1989–90.
- On the national level, master's degrees in science and mathematics rose by 9 percent, and doctor's degrees in science and mathematics rose by 25 percent, between 1985–86 and 1989–90.
- From 1975–76 to 1989–90, bachelor's degrees in computer sciences increased by 385 percent, on the national level. Engineering bachelor's degrees also increased, by 77 percent, and bachelor's degrees in health sciences increased by 9 percent. Bachelor's degrees in agricultural sciences, life sciences, and physical sciences all decreased by over 24 percent during that same period.
- Nonresident aliens were awarded over 50 percent of all doctor's degrees in engineering and mathematics in 1989–90. Master's degrees and doctor's degrees conferred to nonresident aliens in science and mathematics between 1980–81 and 1989–90 increased at a faster rate for nonresident aliens than for U.S. students.

Foreword

Degrees in Science and Mathematics: National Trends and State-By-State Data provides a compilation of recent statistical information on science and mathematics degrees in the United States. This report also describes the larger environment of employment of individuals with science and mathematics degrees, and provides some information on non-U.S. citizens receiving degrees in these areas. The data draw on results of the "Higher Education General Information Survey," "Integrated Postsecondary Education Data System," "Recent College Graduate Survey," and "High School Transcript Study" programs of the National Center for Education Statistics (NCES). This report also contains information provided by the National Science Foundation, the Bureau of the Census, and the Bureau of Labor Statistics.

In working toward the National Education Goal of being first in the world in math and science by the year 2000, and in preparing to meet the challenges of competing in a global economy in the 21st century, members of the education community, and all Americans, need to be kept informed of issues involving science and mathematics. *Degrees in Science and Mathematics: National Trends and State-By-State Data* should be of interest to education researchers and administrators, government officials, the media, the business community, the science community, and the general public.

Emerson J. Elliott
Commissioner of Education Statistics

Acknowledgments

Many people have contributed to the development of this publication. Irene Baden Harwarth originated the idea for this report, designed the tabulations, analyzed the data and wrote the report, under the supervision of Thomas D. Snyder, Chief of the Compilations and Special Studies Branch, and under the general guidance of Jeanne E. Griffith, Associate Commissioner for Data Development. Bill Sonnenberg, Charlene Hoffman, Carol Sue Fromboluti, and Celestine Davis, also of the Compilations and Special Studies Branch, provided statistical advice and support.

The report would not have been possible without the help of Jude Corina of Pinkerton Computer Consultants, who provided the bulk of the computer support. Daryl Harris and Robert Craig of Pinkerton Computer Consultants also provided computer assistance. Jeannette Bernardo and her staff at HCR provided valuable assistance with development of tabulations on nonresident alien graduates, and with various editorial duties.

This publication was reviewed extensively by individuals within and outside the U.S. Department of Education. Susan Ahmed, Nabeel Alsalam, John Burkett, Mary Frase, Debra Gerald, Bernard Greene, Roger Herriot, and Suellen Mauchamer of the National Center for Education Statistics, along with Vance Grant of the Office of Educational Research and Improvement, provided help and advice. Other reviewers included Susan Hill of the National Science Foundation, and Dan Hecker of the Bureau of Labor Statistics. Cover design by Phil Carr.

Executive Summary

It is almost impossible to know if our supply of mathematically and scientifically trained personnel is adequate to meet future needs, because the scientific and technological advances that will shape our lives in the future are unpredictable. Just as fax machines and automatic teller machines (ATMs) have revolutionized our workday and banking habits, other previously unknown inventions are sure to change our lives in the future. Creating, repairing, and operating this equipment will require a workforce with advanced scientific and mathematical training. Yet, how many individuals will be needed, and what level of training they will require (associate, bachelor's, graduate degree) is difficult to predict.

Local planners, policymakers, and business leaders need to prepare for the future using available information. This publication provides information on recent trends of new mathematics and science graduates to help them with that task. The data show where the U.S. currently stands in producing science and mathematics graduates. There is also information on trends in degrees conferred, by state. Data on the salaries and employment status of recent college graduates help us compare the fields of mathematics and science with other areas.

Current status

Bachelor's degrees in science and mathematics declined during the late 1980s, in contrast to increases in the total number of bachelor's degrees. At the same time, there were increases in science and mathematics degrees conferred to nonresident aliens. The 1991 salaries for full-time, full-year workers who were recent college graduates in engineering, health sciences, and mathematics/physical sciences/computer sciences, were generally higher than the average for all recent college graduates. Science and mathematics graduates were also more likely to report having jobs with career potential and jobs that related to their field of study.

Recent trends

National trends in numbers of science and mathematics degrees conferred differ depending on the timeframe studied, field, and the level of degree.

Associate Degrees. Compared with associate degrees conferred in all fields, which increased by 1 percent between 1982–83 and 1989–90, associate degrees in science and mathematics declined by over 9 percent during that time.

Bachelor's Degrees. The total number of bachelor's degrees conferred nationally rose 13 percent between 1975–76 and 1989–90. Bachelor's degrees in science and mathematics fluctuated over this same period. Bachelor's degrees in science rose 21 percent between 1975–76 and 1980–81, and then by a further 15 percent between 1980–81 and 1985–86. Between 1985–86 and 1989–90 the pattern reversed and science degrees dropped 16 percent. Bachelor's degrees in mathematics dropped by 31 percent between 1975–76 and 1980–81, rose 47 percent between 1980–81 and 1985–86 and then declined 10 percent between 1985–86 and 1989–90.

Master's Degrees. Between 1975–76 and 1989–90 the total number of master's degrees conferred rose only 4 percent, compared to an almost 50 percent increase for master's degrees in science. In contrast, during that same time, master's degrees in mathematics dropped by 5 percent.

Doctor's Degrees. Between 1975–76 and 1989–90 doctor's degrees in all fields rose 12 percent, compared to 44 percent for doctor's degrees in science and 7 percent for doctor's degrees in mathematics.

Trends in science fields

Between 1975–76 and 1989–90 the numbers of bachelor's degrees in computer sciences grew at a phenomenal rate (385 percent), engineering and health sciences also grew (77 percent and 9 percent, respectively) but at a slower pace. Bachelor's degrees declined in agricultural sciences (33 percent), life sciences (32 percent), and physical sciences (25 percent). However, in the late 1980s there were declines in the production of bachelor's degrees in every science field, including computer sciences and engineering.

During the late 1980s, master's degrees conferred increased in computer sciences, engineering, and health sciences, while decreasing in agricultural sciences, life sciences, and physical sciences. Doctor's degrees increased in all of the science fields during this time.

Regional differences

No one region dominates the supply of new science and mathematics graduates. According to degree data presented in an earlier NCES publication, *Historical Trends: State Education Facts, 1969 to 1989*, total numbers of bachelor's, master's, and doctor's degrees increased steadily in all the regions throughout the late 1970s and the entire 1980s, and they grew faster in the South and West than in the Northeast and Midwest.

Degrees in Science and Mathematics: National Trends and State-By-State Data shows that in contrast, growth has not been consistent in science and mathematics. The strengths and weaknesses of the regions vary depending on level of degree, and particular field. The most notable regional trends are as follows:

The Midwest. Between 1975–76 and 1989–90, the Midwest led the country each year in the percentage of total bachelor's degrees that were awarded to science graduates (the South had the smallest percentage each year). During that period the Midwest also had the largest increases of all the regions in bachelor's degrees conferred in science, and in master's degrees conferred in mathematics. Out of all the regions, the Midwest had the least amount of growth in master's in science between 1975–76 and 1989–90. In 1989–90 the Midwest was the region that awarded the most doctor's degrees in agricultural sciences, computer sciences, and engineering.

The South. The South had the largest increases in the Nation in awarding doctor's degrees in science and mathematics between 1975–76 and 1989–90 compared to the other regions. In 1989–90 the South led the country in awarding doctor's degrees in life sciences and health sciences.

The Northeast. Between 1975–76 and 1989–90, the Northeast had the smallest growth of all the regions in the numbers of bachelor's degrees awarded in science, but the greatest growth of all the regions in master's in computer science, and was also the region that consistently awarded the largest number of master's degrees in computer science each year. The Northeast had the largest decreases in the country in the numbers of bachelor's and master's awarded in mathematics between 1975–76 and 1989–90.

The West. The West conferred the fewest bachelor's degrees of any of the regions in 1989–90 in mathematics and in all the science fields except for agricultural sciences. In 1989–90, California, with its large population, was often first in the Nation in numbers of degrees conferred in all fields, and at all levels, usually followed by New York.

Foreign students

Half of all the doctor's degrees awarded in 1989–90 in engineering and in mathematics were awarded to nonresident aliens. A third or more of the doctor's degrees, and one quarter of the master's degrees conferred in agricultural sciences, computer sciences, and physical sciences, were earned by nonresident aliens. However, the proportions of associate and bachelor's degrees awarded to nonresident aliens in these subjects, as well as health sciences and life sciences, were much smaller, all under 8 percent.

In 1989–90, overall, 3 percent of all bachelor's degrees, 11 percent of all master's degrees, and 23 percent of all doctor's degrees were awarded to nonresident aliens, compared to 4 percent of bachelor's degrees, 21 percent of master's degrees, and 35 percent of doctor's degrees in science and mathematics.

Notes

In this publication the field "science" is defined as the aggregate of the following fields: agricultural sciences, computer sciences, engineering, health sciences, life sciences, and physical sciences, unless otherwise noted. These fields are sometimes referred to as the "natural sciences and engineering."

In the discussion of the regional and state-by-state data it is important to note that states and regions do not confer degrees; institutions in states and regions do. However, to avoid awkward wording, the word state or the name of the state or the region is used in place of "institutions in the state of..".

Comparisons of data from the Current Population Survey (CPS) in Table 1–3 were analyzed using a weighted regression, and all comparisons were significant at the 95 percent confidence level. CPS data were organized into the following occupational categories:

Computer sciences - computer systems analysts and scientists, operations and systems researchers and analysts, computer programmers, computer equipment operators.

Engineering - aerospace, metallurgical and materials, mining, petroleum, chemical, nuclear, civil, electrical and electronic, industrial, mechanical, engineering related technologists and technicians.

Health sciences - registered nurses, pharmacists, dietitians, physicians assistants, health service occupations.

Physical sciences - physicists and astronomers, chemists (except biochemists), atmospheric and space scientists, geologists and geodesists, physical scientists - other, chemistry teachers (postsecondary), physics teachers (postsecondary), natural science teachers (postsecondary).

Education - teachers except postsecondary.

Social sciences - economists, psychologists, sociologists, social scientists - other, urban planners, social workers.

The use of dashes in the tables, "—", means that no data were reported by institutions, or in some cases, that there were not enough data to show a trend. This is different from "0" which indicates that institutions reported that they did not confer any degrees in that area during that specific period.

For the sake of consistency, the following terms will be used throughout the report to describe major categories of science: "agricultural sciences" to describe subjects categorized under agricultural sciences as well as what is described in some publications as agriculture and natural resources; "computer sciences" to describe subjects categorized under computer sciences as well as what is described in some publications as computer and information sciences; "engineering" to describe subjects categorized under engineering as well as what is described in some publications as engineering and related technologies; "health sciences" to describe subjects categorized under health sciences as well as what is described in some publications as health professions; and "life sciences" to describe subjects categorized under life sciences as well as what is described in some publications as biological sciences. Subcategories for each of these major science categories may be different depending on the source of the data. For more information on these categories, see "Definitions" and the "Guide to Sources" in the back of the publication.

There are several tables in this publication that present degree data from NCES on a year-by-year basis over 7–year and 15–year periods. Data in these tables may occasionally show unusual fluctuations from year to year due to variations in reporting practices or classifications of programs by institutions, or institutions or programs within institutions expanding or closing.

Changes in degree data are reported in these tables in 3–year, 5–year, and 7–year periods for associate degrees, and 5–year and 15–year periods for bachelor's, master's, and doctor's degrees. Providing percentage changes at these intervals does allow for analysis of these data, but it does not make up for fluctuations in reporting from one year to the next.

For more detailed information on individuals with science degrees, refer to the following reports produced by the National Science Foundation: *Characteristics of Recent Science and Engineering Graduates: 1990, Science and Engineering Degrees: 1966–90, and Foreign Participation in U.S. Academic Science and Engineering: 1991.*

Notes table. — Definitions of science, science fields.

Definitions of science		Definitions of science fields		
Degrees in Science and Mathematics: National Trends and State-By-State Data	National Science Foundation	*Degrees in Science and Mathematics: National Trends and State-By-State Data*	Integrated Postsecondary Education Data System	Recent College Graduates Survey
Science	Science	Agricultural sciences	Agriculture and natural resources	
Agricultural sciences Computer sciences Engineering Health sciences Life sciences Physical sciences	Engineering Physical sciences Earth, atmospheric and ocean sciences Mathematical/computer sciences Biological/agricultural sciences Psychology Social sciences	Computer sciences	Computer and information sciences	Math/computer sciences/physical sciences
		Engineering	Engineering Engineering technologies	Engineering
		Health sciences	Health sciences	Health professions
		Life sciences	Life sciences	Biological sciences
		Physical sciences	Physical sciences	Math/computer sciences/physical sciences

Contents

Tables

Charts

Introduction

It is almost impossible to know if our supply of mathematically and scientifically trained personnel is adequate to meet future needs, because the scientific and technological advances that will shape our lives in the future are unpredictable. Just as fax machines and automatic teller machines (ATMs) have revolutionized our workday and banking habits, other previously unknown inventions are sure to change our lives in the future. Creating, repairing, and operating this equipment will require a workforce with advanced scientific and mathematical training. Yet, how many individuals will be needed, and what level of training they will require (associate, bachelor's, graduate degree) is difficult to predict.

Local planners, policymakers, and business leaders need to prepare for the future using available information. This publication provides information on recent trends of new mathematics and science graduates to help them with that task. The data show where the U.S. currently stands in producing new science and mathematics graduates. There is also information on trends in degrees conferred, by state. Data on the salaries and employment status of recent college graduates help us compare the fields of mathematics and science with other areas.

Issues in Demand for Graduates in Science and Mathematics

A controversy has evolved about the job market for scientifically and technically trained personnel in the U.S. One article in *The Chronicle of Higher Education* described rosy job prospects for new graduates in science and mathematics fields:

> *"Unlike most of their classmates, who are fighting for jobs in the troubled economy, college seniors in engineering, chemistry, computer science, and various health-care fields are being inundated with job offers . . . such fields as health care, engineering, and computer technology had more job openings than there were graduates to fill them."*[1]

On the other hand, there is also evidence that the supply of new graduates in science and mathematics may be greater than the demand for their services. Recent congressional hearings have investigated complaints from scientists and engineers of few job offers and low salaries offered to doctor's degree recipients in science fields. *Science* magazine recently reported on the difficulties facing physics majors with doctor's degrees:

> *"Take last month's 1400–attendee International High-Energy Physics meeting in Dallas. Nearly every young physicist approached by* Science *was job-hunting. Most said they were getting desperate. . . . The pool is growing, the demand shrinking, and the pipeline of physics clogging."*[2]

These conflicting viewpoints lead to speculation: Is this just a small, overly specialized group struggling with changes in industry and research? Are these articles anecdotal, or are there solid data to back up these claims?

Degrees in Science and Mathematics, National Trends and State-By-State Data

This publication provides information on the issues surrounding supply of and demand for new graduates in science and mathematics on the national level by looking at indicators of demand for these graduates; national trends over a 15–year period in the awarding of associate, bachelor's, master's, and doctor's degrees in science and mathematics; and the "pipeline" of students studying science and mathematics. Regional and state-by-state figures for science and mathematics degree recipients will also be provided, as well as information on foreign students or "nonresident aliens" receiving degrees in science and mathematics.

Information is presented on the market for new graduates in science and mathematics by discussing the numbers of scientists and mathematicians currently in the workforce, the earnings of employed scientists and mathematicians, and employment rates and salaries for recent college graduates in science and mathematics. Data in this publication on degrees awarded in science and mathematics are organized in tables by associate, bachelor's, master's, and doctor's degrees in all sciences, and then specifically in the following subject areas: agricultural sciences, computer sciences, engineering, health sciences, life sciences, physical sciences, and mathematics. An additional section provides national

[1] "For Some Graduating Seniors, Job Offers Abound; for Most, the Outlook is Gloomy." *The Chronicle of Higher Education*, May 20, 1992, p. A28.

[2] "Physics Famine: A Frenzied Search for Job Stability." *Science*, September 18, 1992, Vol.257, p. 1726.

data on nonresident aliens receiving science and mathematics degrees during the 1980s.

Chapter 1 of this report focuses on labor market outcomes which serve as indicators of marketplace demand for new graduates in science and mathematics. Chapter 2 discusses the science and mathematics education pipeline and the national trends in science and mathematics degrees. Chapter 3 presents analysis of data on regional and state-by-state trends in mathematics degrees. Chapter 4 provides analysis of data on regional and state-by-state trends in science degrees, and Chapter 5 provides a closer look at these science degrees by categorizing them into six subject fields: agricultural sciences, computer sciences, engineering, health sciences, life sciences, and physical sciences. Chapters 3 through 5 provide tables showing data by degree level, on a regional and state-by-state basis. Chapter 6 analyzes data on nonresident aliens who were awarded degrees in the United States, and on the doctor's level what nonresident aliens have identified as their postgraduate plans. The appendices provide a listing of definitions and sources for information in this publication.

In the words of one economist from the Bureau of Labor Statistics: "Our Nation's economic progress and general well-being depend in considerable measure on the work of scientists, engineers, and technicians. These men and women contribute to the development of new products, improvements in productivity, enhanced defense capabilities, environmental protection, and advances in communications and health care."[3] One of the National Education Goals agreed upon by then-President George Bush, and the Nation's Governors led by then-Governor Bill Clinton, in an historic meeting on education in 1989, was the following: "By the year 2000, U.S. students will be first in the world in science and mathematics achievement." This goal recognizes the important role science and mathematics will play in the next century. The goal reflects the conviction that there is a linkage between science and mathematics achievement and economic competitiveness. However, science and mathematics are not only connected to economic competitiveness on the international level, but on the national, regional, and state levels as well.

The data presented in this publication will be especially useful to state policymakers concerned with improving their states' economic performance as well as educational opportunities. Information on mathematically and scientifically trained personnel presented on a state-by-state and regional basis can be valuable to both business and government leaders as executives in industry choose sites of future factories and corporate offices and evaluate the availability of scientific and technical personnel in different states. In addition to offering new data to the public in these important areas of science and mathematics, this publication will raise questions for future research.

[3] Braddock, Douglas J., "Scientific and Technical employment, 1990–2005." *Monthly Labor Review*, February, 1992, p.28.

1. Demand For Graduates in Science and Mathematics: A Few Indicators

There are many issues involved in assessing demand for scientifically and mathematically trained personnel in the U.S. Among the determining factors are: the employment of recent college graduates, the staffing requirements of industry, the current status of scientists and mathematicians in the workforce, the rates of retention and turnover, the needs of the defense/aerospace industry, the demands of the global marketplace, and the overall behavior of the economy.

According to the National Science Board, private industry in the U.S. employed approximately 2 million scientists and engineers in 1989. These scientists and engineers made up 2.4 percent of the private labor force.[1] This was up from 1.3 million scientists and engineers, or 2.1 percent of the private labor force, in 1980.[2] Smaller numbers of scientists and engineers work for federal, state, and local government agencies.[3]

Numbers of positions for scientists and engineers have grown at almost twice the rate for all workers in private industry.[4] Of special interest is the phenomenal growth in the computer and data processing sector during the 1980s. Although the **numbers** in this industry are small, the **rate of growth** has been very high. At an average growth of 13.1 percent per year between 1980 and 1989, this industry had a 205 percent increase in the number of positions, rising from 41,000 in 1980 to 125,000 in 1989.[5] From 1967 to 1989 there was a decline in positions for scientists and engineers in manufacturing and an increase in positions in nonmanufacturing areas such as business and financial services.[6] The defense buildup of the early 1980s created opportunities for science and engineering graduates and, in upcoming chapters in this report, the data show increases in computer science and engineering degrees during portions of the 1980s.

According to various mathematical models of demand,[7] the numbers of scientists and engineers needed in the 1990s are projected by NSF to rise anywhere from about 14 percent to a possible 27 percent. The NSF considers prediction of demand for scientists as a long-term process, affected more by factors such as a slightly higher rate of retirement for those currently in the field than by minor short-term adjustments due to the business cycle.[8] The Bureau of Labor Statistics (BLS) (see table 1–1), in its *Occupational Outlook Handbook*, has also predicted a growth rate of "faster than average" through the year 2005 for most occupations requiring some type of background in science or mathematics. Regardless of the model, each scenario depends on long-term estimates of how the U.S. economy handles the expected shift from scientific research and development in defense to increased research and development in the production of non-defense goods. Other uncertain factors in the models include the volume of U.S. imports and exports.

Indicators of Demand for Science and Mathematics Graduates

Indicators can be used to evaluate the absorption of scientists and engineers into the economy. These indicators can help us assess the relative supply and demand in different time periods or across different occupations or fields of study. Some indicators of workforce demand are:

- salaries of recent college graduates,
- employment of recent college graduates,
- unemployment rates of recent college graduates,
- underemployment rates of recent college graduates, and
- earnings of scientists and engineers currently in the workforce.

Salaries of Recent College Graduates

Table 1–2 presents data from the 1991 Survey of Recent College Graduates (RCG), which show that one year after earning their bachelor's degrees, graduates in health sciences, engineering, and mathematics/computer sciences/physical sciences employed full-time all had mean salaries that were higher than the average for all graduates employed full-time. Those who earned bachelor's degrees in life sciences had average salaries that were lower

---

[1] National Science Board. "Science and Engineering Workforce" *Science & Engineering Indicators, 1991.* (Washington, D.C.: U.S. Government Printing Office, 1991) pp. 67, 267.

[2] Ibid. p. 267.

[3] The Bureau of Labor Statistics estimates the 1990 count of engineers in federal, state, and local government employ was approximately 200,000.

[4] National Science Board. p. 67.

[5] Ibid. p. 70.

[6] Ibid. pp. 67, 69.

[7] Ibid. pp. 80–81.

[8] Ibid. pp. 76, 79.

than the average salary for all bachelor's degree recipients employed full-time.[9]

Recent bachelor's degree recipients in social sciences, public affairs/social services, psychology, education, and the humanities had average salaries that were lower than the average for all recent bachelor's degree recipients. Those who earned bachelor's degrees in business and management reported salaries that were higher than the average for all recent bachelor's degree recipients.[10]

Table 1–2 also shows that from 1987 to 1991, when adjusted for inflation, the average salaries going to recent bachelor's degree recipients in the categories of health sciences and mathematics/computer sciences/physical sciences increased. However, during that same time, average salaries for recent bachelor's degree recipients in engineering and life sciences did not change when tested for statistical significance.

Between 1987 and 1991 average salaries for recent bachelor's degree recipients in business and management, education, humanities, psychology, public affairs/social services, and social sciences were stable when tested statistically. Overall, the salary data over time seem to indicate a demand for individuals with certain types of science and mathematics majors.[11]

Employment of Recent College Graduates

Recent college graduates with bachelor's degrees in engineering and health sciences had full-time employment rates that were higher than that for all majors. Engineering and health majors were more likely to be employed in a field related to their study. They also were more likely to state that their jobs had some career potential; that is, the job provided a future with the company and allowed for advancement and promotions (for more details, see table 1–2). Graduates with bachelor's degrees in mathematics/computer sciences/physical sciences had the same rate of full-time employment as graduates in all majors; yet they were more likely to say that they had jobs related to their field of study, and with some career potential. Life sciences majors had lower full-time employment rates, and were less likely to have jobs with career potential, than recent bachelor's degree recipients in all majors.

Table 1–2 shows that recent bachelor's degree recipients in business and management and in education had full-time employment rates that were higher than that for all majors. It is notable that bachelor's degree recipients with majors in business and management, humanities, psychology, public affairs, and the social sciences were more likely than the average graduate to feel their jobs did not require a four-year degree.

Underemployment and Unemployment of Recent College Graduates

Data in table 1–2 show that among those employed full-time, recent college bachelor's degree recipients in life sciences, engineering, health sciences, and mathematics/computer sciences/physical sciences had lower rates of "underemployment" than all graduates. Underemployment is the term used to describe full-time jobs which did not require a college degree such as sales, service, administrative support, crafts, operators, and laborers. Recent graduates with bachelor's degrees in business and management, humanities, public affairs/social service, and social sciences were more likely to be underemployed than the average recent college graduate.

Health sciences majors had a lower unemployment rate than all graduates. Bachelor's degree recipients in engineering, life sciences, and mathematics/computer sciences/physical sciences had average percentages of unemployment among their ranks. Among bachelor's degree recipients in other selected fields, humanities majors had higher percentages of unemployed graduates than the average, while education majors had a lower percentage of unemployed graduates. For more detailed data on recent college graduates in science see *Characteristics of Recent Science and Engineering Graduates: 1990* by the National Science Foundation (NSF).

Earnings of Workers in Science Occupations and Other Occupations

For 1970 through 1990, workers in the occupations of engineering, computer sciences, or physical sci-

---

[9] All differences cited based on the Recent College Graduate (RCG) data in Table 1–2 are significant at the 5 percent significance level. Differences between each field and the total were tested using a t-test in which the standard error was adjusted for the covariance between the subgroup (field) and the total. The exact formula is

$$t = \frac{\overline{X_s} - \overline{X_T}}{[se_S^2 + se_T^2 - 2(p)se_S^2]^{1/2}}$$

where S = subgroup, T = total, X = mean, se = standard error, and p = the proportion of the total group contained in the subgroup. A Bonferroni adjustment was applied to adjust for multiple comparisons; the comparison size was 10, accounting for the 10 fields of study.

[10] National Center for Education Statistics. *Occupational and Educational Outcomes of 1989–90 Recent College Graduates 1 Year After Graduation: 1991* (Washington, D.C.: U.S. Department of Education, 1993) p. 13.

[11] National Center for Education Statistics. *Occupational and Educational Outcomes of 1985–86 Bachelor's Degree Recipients 1 Year After Graduation: 1987*, (Washington, D.C.: U.S. Department of Education, 1989) p. 3.

ences had higher median earnings than the median for all full-time, full-year workers with a bachelor's degree (table 1–3). Individuals working in health sciences had median earnings that were lower than the median earnings of bachelor's degree recipients, although this gap has been narrowing during the late 1980s.

Median earnings of educators were lower than the median earnings for individuals in all occupations for each year from 1970 through 1990 (chart 1). Individuals working in the social sciences also had lower earnings than the median for these years, except for 1975, 1976, 1981, and 1984.[12]

Chart 1 — Median earnings of full-time, full-year workers with at least a bachelor's degree, for selected occupations, in constant 1990 dollars: 1970–1990

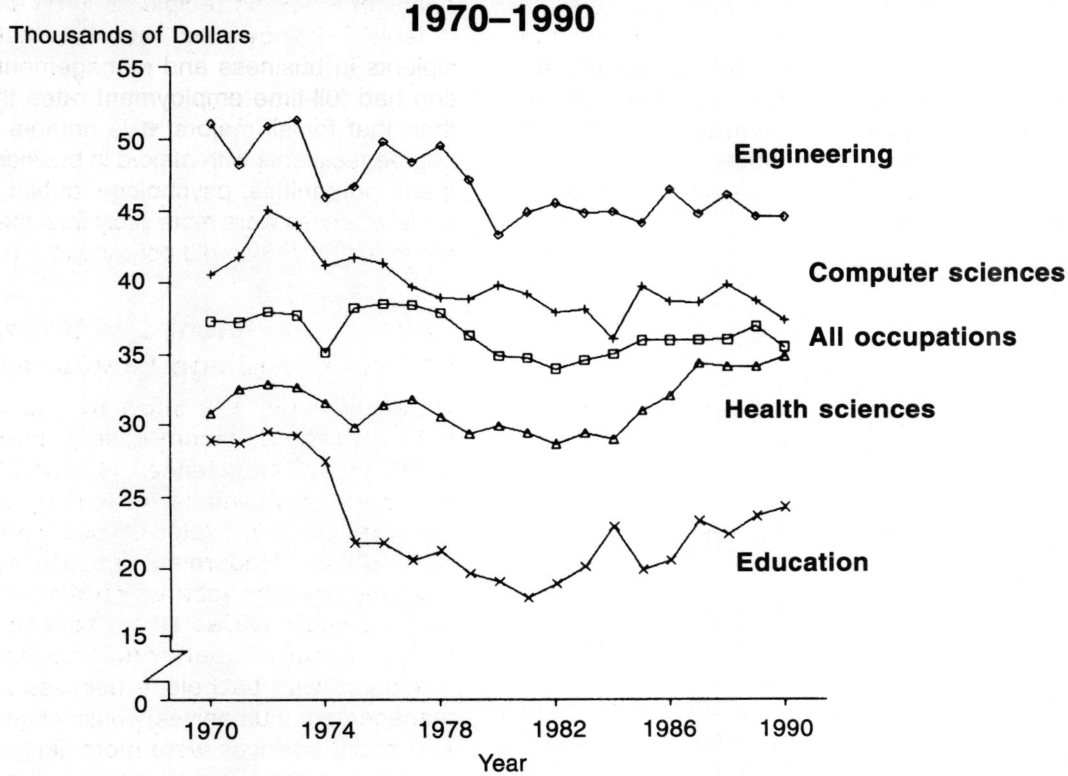

Source: U.S. Bureau of the Census, March Current Population Survey, 1971 through 1991, unpublished data.

Summary

Data presented in this chapter indicate a general pattern of higher salaries and less underemployment for recent science and mathematics graduates. Average earnings for all workers in the fields of engineering, computer sciences, and physical sciences were higher than the earnings for individuals in all occupations. Overall, these data present some evidence that there may be better labor market opportunities for scientifically and mathematically trained personnel.

Further study is necessary to describe the dynamics between supply and demand for new graduates in science and mathematics compared to new graduates in other academic areas.

[12] Life sciences and mathematics were not included because there were too few sample cases (under 75,000 weighted cases) for a reliable estimate. Agricultural sciences data were not consistent with other data. Data from the Current Population Survey (CPS) in Table 1–3 were analyzed using a weighted regression (weight=inverse of standard error squared). The dependent variable was the difference between occupation mean and overall mean. The independent variable was centered year (year-11). Year was coded as 1 to 21. Linear and quadratic regressions were run. A zero intercept in the linear model indicates no difference on average between the specific occupation and the total. A zero slope indicates a constant difference between the specific occupation and the total over time. A nonzero slope indicates that the differences between the specific group and the total are changing over time. Because of correlations in the CPS data in consecutive years, the analysis was repeated twice, using half of the data at one time (even years or odd years). Results were similar in the two analyses.

Table 1-1.— Bureau of Labor Statistics employment outlook for selected occupations: 1990 through 2005

Occupation	1990 employment (in thousands)	Outlook through the year 2005
Total labor force	125,000	151,000
Selected science and engineering occupations		Growth rate 1990 to 2005 (average = 21 %)
Engineers	1,519	faster than average
Electrical	426	faster than average
Mechanical	233	average
Civil	198	faster than average
Industrial	135	average
Aerospace	73	average
Chemical	48	slower than average
Nuclear	18	slower than average
Metallurgical	18	average
Petroleum	17	slower than average
Mining	4	slower than average
Engineering, science, and data processing managers	315	faster than average
Actuaries	13	faster than average
Computer systems analysts	463	faster than average
Agricultural scientists	25	faster than average
Biological scientists	62	faster than average
Chemists	83	average
Physicists and astronomers	20	slower than average
Occupational therapists	36	faster than average
Pharmacists	169	faster than average
Physical therapists	88	faster than average
Physician assistants	53	faster than average
Recreational therapists	32	faster than average
Registered nurses	1,727	faster than average
Respiratory therapists	60	faster than average
Speech-language pathologists and audiologists	68	faster than average
Other selected fields		
Social scientists	2,240	faster than average
Psychologists	125	faster than average
Kindergarten and elementary school teachers	1,520	average
Secondary school teachers	1,280	faster than average

Source: U.S. Department of Labor, Bureau of Labor Statistics, *Occupational Outlook Handbook* 1992-93 Edition, various chapters.

THIS PAGE INTENTIONALLY LEFT BLANK

Table 1-2.—Employment, school enrollment, and salaries of recent bachelor's degree recipients, by field, 1991, and salaries of recent bachelor's degree recipients, by field, 1987

Field of study	Full-time employed		All graduates		Full-time and part-time employed		
	Average annual salary	Percent under-employment[1]	Percent employed full-time	Percent employed part-time	Percent job related to field of study	Percent some career potential of job	Percent four-year degree not required for job
All majors	$23,632	23	74	11	76	79	44
Science and mathematics majors							
Life sciences	\-\ 21,051	\-\ 16	\-\ 51	\=\ 12	\=\ 73	\-\ 67	\=\ 42
Engineering	\+\ 30,933	\-\ 7	\+\ 85	\-\ 3	\+\ 89	\+\ 90	\-\ 19
Health sciences	\+\ 31,455	\-\ 4	\+\ 81	\=\ 11	\+\ 95	\+\ 92	\=\ 49
Mathematics/computer sciences/physical sciences	\+\ 27,156	\-\ 10	\=\ 71	\-\ 8	\+\ 86	\+\ 85	\-\ 33
Other selected majors							
Business and management	\+\ 24,727	\+\ 28	\+\ 83	\-\ 6	\+\ 81	\+\ 83	\+\ 47
Education	\-\ 19,110	\-\ 11	\+\ 77	\+\ 15	\+\ 87	\+\ 84	\-\ 24
Humanities	\-\ 19,059	\+\ 33	\-\ 59	\+\ 19	\-\ 57	\-\ 66	\+\ 57
Psychology	\-\ 19,154	\=\ 28	\-\ 60	\=\ 14	\-\ 65	\-\ 69	\+\ 53
Public affairs/social service	\-\ 20,801	\+\ 35	\=\ 77	\=\ 11	\=\ 71	\-\ 71	\+\ 52
Social sciences	\-\ 22,213	\+\ 31	\-\ 68	\=\ 12	\-\ 53	\-\ 72	\+\ 52

Field of study	Percent of all graduates enrolled full-time in further education	Percent of those in labor force unemployed[2]	Average annual salary of full-time employed in 1991 constant dollars		
			1989–90 graduates in 1991	1985–86 graduates in 1987	Percent change
All majors	17	4.5	$23,632	$24,399	−3.1
Science and mathematics majors					
Life sciences	\=\ 17	\=\ 4.2	21,051	19,632	7.2
Engineering	\=\ 18	\=\ 3.4	30,933	31,892	−3.0
Health sciences	\=\ 16	\-\ 1.0	31,455	27,045	16.3
Mathematics/computer sciences/physical sciences	\=\ 17	\=\ 5.1	27,156	21,193	28.1
Other selected majors					
Business and management	\-\ 13	\=\ 5.2	24,727	25,292	-2.2
Education	\+\ 27	\-\ 2.2	19,110	18,897	1.1
Humanities	\=\ 17	\+\ 6.2	19,059	19,449	-2.0
Psychology	\=\ 19	\=\ 6.7	19,154	20,760	-7.7
Public affairs/social service	\-\ 13	\=\ 4.6	20,801	21,193	-1.8
Social sciences	\=\ 16	\=\ 4.9	22,213	24,379	-8.9

[1]Underemployment refers to graduates who are employed full-time in the jobs of sales, service, administrative support, crafts, operators, and laborers, and who indicated a college degree was not required for job.

[2]Unemployment refers to not working and both looking for work and available for work.

\-\ Significantly lower than average for total graduates.

\=\ No significant difference from the average for total graduates.

\+\ Significantly higher than average for total graduates.

Note: All differences cited are significant at the 95 percent confidence level. Differences between each field and the total were tested using a t-test in which the standard error was adjusted for the covariance between the subgroup (field) and the total. The exact formula is provided in footnote 9 to this chapter. A Bonferroni adjustment was applied for multiple comparisons; the comparison size was 10, accounting for the 10 fields of study. For more detailed data on specific fields, refer to "Characteristics of Recent Science and Engineering Graduates: 1990" by the National Science Foundation (NSF). se=standard error.

Source: National Center for Education Statistics, *Occupational and Educational Outcomes of 1989-90 Bachelor's Degree Recipients 1 Year After Graduation: 1991*, pages 7, 12, 13, 16, 17, 18, and tables A-2, A-3, A-5, A-6, A-7, A-8, A-9, and National Center for Education Statistics, *Occupational and Educational Outcomes of 1985-86 Bachelor's Degree Recipients 1 Year After Graduation: 1987*, page 37.

Table 1-2 (cont.)—Standard errors for employment, school enrollment, and salaries of recent bachelor's degree recipients, by field, in 1991, and salaries of recent bachelor's degree recipients, by field, 1987

Field of study	Full-time employed				All graduates				Full-time and part-time employed					
	Average annual salary of full-time employed	se	Percent under-employment[1]	se	Percent employed full-time	se	Percent employed part-time	se	Percent job related to field of study	se	Percent some career potential of job	se	Percent four-year degree not required for job	se
All majors	$23,632	180.0	23	0.4	74	0.4	11	0.3	76	0.5	79	0.4	44	0.6
Science and mathematics majors														
Life sciences	21,051	414.0	16	2.0	51	1.7	12	1.0	73	1.8	67	2.3	42	2.0
Engineering	30,933	392.0	7	0.8	85	1.0	3	0.5	89	1.2	90	0.7	19	1.5
Health sciences	31,455	859.0	4	0.6	81	1.4	11	1.3	95	0.8	92	0.9	49	3.5
Mathematics/computer sciences/physical sciences	27,156	399.0	10	1.3	71	1.3	8	0.6	86	1.1	85	1.1	33	2.7
Other selected majors														
Business and management	24,727	324.0	28	0.9	83	0.6	6	0.3	81	0.7	83	0.6	47	1.0
Education	19,110	137.0	11	0.8	77	0.7	15	0.7	87	0.7	84	0.8	24	1.0
Humanities	19,059	340.0	33	1.4	59	1.3	19	1.2	57	1.7	66	1.6	57	1.4
Psychology	19,154	314.0	28	1.8	60	1.6	14	1.3	65	1.5	69	1.5	53	1.9
Public affairs/social service	20,801	466.0	35	2.1	77	2.1	11	1.2	71	2.2	71	2.2	52	2.2
Social sciences	22,213	327.0	33	1.2	68	0.9	12	0.6	53	1.3	72	1.0	52	1.1

Field of study	Percent of all graduates enrolled full-time in further education[2]	se	Percent of those in labor force unemployed[2]	se	Average annual salary of 1985–86 graduates full-time employed in 1987 in 1991 constant dollars	se
All majors	17	0.4	4.5	0.2	$24,399	162.2
Science and mathematics majors						
Life sciences	17	1.4	4.2	1.1	19,632	529.8
Engineering	18	1.3	3.4	0.5	31,892	263.4
Health sciences	16	1.5	1.0	0.3	27,045	267.5
Mathematics/computer sciences/physical sciences	17	1.1	5.1	0.7	21,193	289.5
Other selected majors						
Business and management	13	0.7	5.2	0.3	25,292	306.3
Education	27	0.8	2.2	0.3	18,897	173.3
Humanities	17	1.3	6.2	0.6	19,449	256.1
Psychology	19	1.3	6.7	1.0	20,760	524.1
Public affairs/social service	13	1.2	4.6	0.9	21,193	456.3
Social sciences	16	0.6	4.9	0.6	24,379	709.2

Table 1-3.—Median earnings of full-time, full-year workers with at least a bachelor's degree, for selected occupations, in constant 1990 dollars: 1970 to 1990

	1970	1971	1972	1973	1974	1975	1976	1977	1978	1979	1980	1981	1982	1983	1984
All occupations	$37,289	$37,219	$37,935	$37,712	$35,097	$38,264	$38,547	$38,455	$37,886	$36,303	$34,925	$34,764	$34,018	$34,639	$35,044
Computer sciences	40,585	41,822	45,069	44,020	41,265	41,764	41,404	39,715	38,949	38,905	39,839	39,261	37,943	38,176	36,122
Engineering	51,053	48,195	50,909	51,341	45,957	46,712	49,821	48,398	49,562	47,207	43,344	44,952	45,594	44,907	45,025
Health sciences	30,814	32,557	32,918	32,717	31,591	29,777	31,438	31,846	30,578	29,386	29,928	29,419	28,672	29,403	29,022
Physical sciences	45,451	44,727	46,239	49,611	43,510	45,467	44,667	44,476	45,454	43,608	38,423	38,449	42,728	42,413	42,424
Other selected occupations															
Education	28,898	28,724	29,516	29,225	27,481	21,787	21,786	20,599	21,227	19,608	19,077	17,915	18,940	20,125	22,975
Social sciences	33,222	34,231	37,389	34,341	32,998	41,062	44,695	36,233	34,832	32,388	34,373	36,603	28,264	26,787	35,090
Standard errors															
All occupations	129.5	123.3	94.1	170.1	163.1	208.1	208.5	279.9	261.6	236.7	209.4	362.5	278.8	242.0	396.3
Computer sciences	1,102.2	683.2	901.0	1,067.6	1,084.1	894.9	845.7	765.1	1,724.2	1,009.1	927.8	1,220.0	1,059.4	1,262.5	1,379.3
Engineering	364.6	468.5	487.7	506.2	520.0	431.5	442.0	596.6	564.2	567.4	592.4	524.9	938.2	806.0	479.1
Health sciences	521.6	426.8	580.3	471.7	555.8	379.6	427.4	699.9	528.5	578.8	630.8	463.2	546.6	688.6	816.5
Physical sciences	1,270.3	840.6	891.9	1,020.4	1,035.5	1,187.8	1,884.4	757.1	1,563.3	1,708.3	1,830.7	1,555.8	1,569.3	1,741.5	1,141.9
Other selected occupations															
Education	124.3	125.2	157.4	157.3	178.7	519.2	427.3	696.6	585.1	584.1	568.3	868.9	845.6	916.2	1,022.1
Social sciences	648.6	601.3	667.9	509.9	700.0	1,466.5	3,056.5	1,975.5	2,306.7	1,607.8	1,533.3	1,774.5	3,207.3	1,594.3	2,015.9

	1985	1986	1987	1988	1989	1990
All occupations	$36,002	$36,001	$36,019	$36,076	$37,002	$35,570
Computer sciences	39,810	38,714	38,693	39,950	38,796	37,486
Engineering	44,218	46,561	44,864	46,211	44,698	44,683
Health sciences	31,072	32,129	34,403	34,209	34,167	34,903
Pysical sciences	47,547	47,294	49,831	44,197	42,793	38,428
Other selected occupations						
Education	19,917	20,554	23,366	22,380	23,721	24,349
Social sciences	24,921	32,558	26,492	28,696	28,849	27,487
Standard errors						
All occupations	401.4	211.5	218.2	389.2	311.0	256.8
Computer sciences	1,167.2	944.6	1,104.8	888.5	617.7	1,119.5
Engineering	671.6	961.8	1,091.8	583.9	899.9	1,177.6
Health sciences	531.5	761.9	1,142.2	507.3	878.2	868.1
Physical sciences	2,664.3	2,541.7	2,577.5	3,718.9	2,044.2	3,040.4
Other selected occupations						
Education	899.9	978.1	1,379.4	1,188.3	1,006.9	1,288.2
Social sciences	3,500.1	2,177.2	1,241.9	1,053.3	1,825.4	2,026.6

NOTE.—Occupations discussed in this table were categorized as follows: Computer sciences - computer systems analysts and scientists, operations and systems researchers and analysts, computer programmers, computer equipment operators. Engineering - aerospace, metallurgical and materials, mining, petroleum, chemical, nuclear, civil, electrical and electronic, industrial, mechanical, engineering related technologists and technicians. Health sciences - registered nurses, pharmacists, dieticians, physicians assistants, health service occupations. Physical sciences - physicists and astronomers, chemists (except biochemists), atmospheric and space scientists, geologists and geodesists, physical scientists - other, chemistry teachers (postsecondary), natural science teachers (postsecondary). Education - teachers, except postsecondary. Social sciences - economists, psychologists, sociologists, social sciences - other, urban planners, social workers.

Source: U.S. Bureau of the Census, March Current Population Survey, 1971 through 1991, unpublished data.

2. Supply of Graduates in Science and Mathematics: Pipeline Issues and National Trends

The word "pipeline" is used to illustrate the flow of individuals through the series of steps leading to completion of a particular field of study. Some issues involving the "pipeline" for scientists and mathematicians include student interest in science and mathematics, student coursework, plans to attend college and obtain a degree in science or mathematics, and plans to pursue a career in science or mathematics. New graduates are an important source for scientists and technically trained workers. Analyzing this pipeline, and national trends of degrees conferred in science and mathematics, provides some indication as to the flow of future workers into science and mathematics fields.

Pipeline Issues

Elementary and Secondary Students

The flow of students through the pipeline narrows in secondary school. According to one report: "Nearly 30 percent of all seventh grade students . . . expressed a preference for an SME (science, mathematics, or engineering) career but these percentages declined steadily throughout the remaining middle and high school years . . . By the 12th grade, fewer than 1 in 4 male students and only 1 in 10 female students expressed an interest in an SME career."[1] Influences on students' choice of a career in science and mathematics include encouragement by parents, parental resources, gender, and persistence in mathematics.[2]

The 1985–86 National Survey of Science and Mathematics Education found that in elementary school, teachers spent the most time on reading instruction, less time on mathematics, and much less time on science instruction.[3] In middle schools and high schools, researchers found significant changes have occurred in course patterns. In the late 1970s and early 1980s: "...states and local education agencies increased their graduation requirements to encourage achievement of academic excellence."[4] The number of credits earned in science and mathematics rose in response to these requirements. A study of high school transcripts comparing 1982 and 1990 high school graduates found that the average number of credits earned in mathematics went up from 2.5 in 1982 to 3.1 in 1990. This also happened in science, where the average number of credits earned by high school graduates rose from 2.2 to 2.8.[5] Between 1981 and 1992, the rate of 11th and 12th graders taking Advanced Placement Examinations in calculus rose from 4 per 1,000 students to 16 per 1,000 students. Science exam takers (biology, chemistry, and physics) rose from 4 to 15 exam takers per 1,000 during the same period.[6]

Despite data showing that more students are taking more science and mathematics courses, and are voluntarily taking Advanced Placement Examinations in these fields, proficiency has not improved dramatically. Student performance data show modest changes in proficiency over a 17–year period. However, some improvements have been registered in the last few years. Proficiency in mathematics for 9– and 13–year-olds improved slightly between 1973 and 1990. However, over the same period, scores for 17–year-olds showed no improvement.[7] Data regarding science achievement reveal no change in proficiency of 9–year-olds and 13–year-olds, and lower science achievement among 17–year-olds in 1990 compared with 1970. However, test scores in both science and mathematics for 17–year-olds, while dropping between 1973 and 1982, rose again between 1982 and 1990.[8] A 1992 mathematics assessment showed an increase in the average proficiency scores for students in grades 4, 8, and 12 after 1990.[9]

Higher Education Students

Increasing college enrollments are fueled by students who are older than the average college stu-

---

1 National Science Board. *Science & Engineering Indicators, 1991*, (Washington, D.C.: U.S. Government Printing Office, 1992) p. 24.

2 Ibid.

3 Ibid. p. 27.

4 Ibid. p. 25.

5 National Center for Education Statistics. *Comparative Data on Credits Earned and Demographics for 1990, 1987, and 1982 High School Graduates.* (Washington, D.C.: National Center for Education Statistics, 1993) p. A-110.

6 National Education Goals Panel. *The National Education Goals Report: Building a Nation of Learners.* (Washington, D.C.: U.S. Government Printing Office, 1992) p. 30.

7 National Center for Education Statistics. *The Condition of Education, 1992* (Washington, D.C.: U.S. Government Printing Office, 1992) p. 46.

8 Ibid. pp. 46, 48.

9 National Center for Education Statistics. *NAEP 1992: Mathematics Report Card for the Nation and the States.* (Washington, D.C.: U.S. Government Printing Office, 1993) p. 1.

dent, females, part-time students, those returning to school after a break in their studies, and those attending a 2–year institution. Such students have shown lower than average tendencies to pursue degrees in science and mathematics in the past.[10] While total enrollment in 4–year institutions increased almost 15 percent from fall 1976 through fall 1988, there were substantial declines in the numbers of students whose major fields of study were life sciences (down 22 percent) and physical sciences (down 16 percent). The number of engineering majors peaked in 1984 and subsequently declined.[11]

In the fall of 1989, 58 percent of all higher education students were under 25, 24 percent were 25 to 34, and 18 percent were over 35. Broken down by major field of study, more than 66 percent of students majoring in agricultural sciences, engineering, and life sciences were reported to be under 25.[12] Data from fall 1988 show that 76 percent of all undergraduates attended college on a full-time basis and 24 percent attended on a part-time basis. However, in engineering, life sciences, mathematics, and physical sciences over 84 percent of undergraduates study full-time and less than 16 percent study part-time. Women make up 53 percent of all the undergraduates, but only 15 percent of undergraduates studying engineering, 46 percent studying mathematics, and 31 percent studying physical sciences. In life sciences, however, women accounted for 52 percent of the undergraduates studying in this field, nearly equalling overall enrollment proportions.[13]

National Trends in Science and Mathematics Degrees

Higher education degrees are awarded at the conclusion of the science and mathematics education pipeline. The numbers of degrees conferred in science and mathematics over the 15–year period from 1975–76 through 1989–90 indicate how the pipeline is reacting to the demand for individuals with science and mathematical training who will be available to meet the needs of industry, academia, government, and the military in the upcoming decades.

Overall, associate and bachelor's degrees conferred in science peaked during the mid-1980s and then declined. The number of master's degrees conferred in science rose every year between 1975–76 and 1989–90. The number of doctor's degrees in science fluctuated in the late 1970s, and rose during the 1980s. In mathematics, the numbers of bachelor's and master's degrees conferred dropped between 1975–76 and 1980–81 and rose in the 1980s. There was a decrease in the number of bachelor's and master's degrees in mathematics over the entire period, and an increase in doctor's degrees. The following is a look at the national trends in degrees conferred in science and mathematics, compared to degrees overall, and degrees in non-science and mathematics fields, by degree level.

Associate Degrees

While associate degrees conferred in all fields increased by 1 percent between 1982–83 and 1989–90, degrees awarded in mathematics and science declined. (Note: 1982–83 is the first year for which the field of study data are consistent for associate degrees.) As can be seen in table 2–1, associate degrees in mathematics (of which few were awarded) declined by 2 percent between 1982–83 and 1989–90, while associates in science declined by over 9 percent during that same period.

Bachelor's Degrees

The number of bachelor's degrees conferred in all fields rose from 926,000 in 1975–76 to 1,050,000 in 1989–90, an increase of 13 percent (see table 2–2). Increases occurred in some science fields, but most of this increase was in non-science and mathematics fields. Computer sciences had the largest *percentage* increase of the fields shown, 385 percent, between 1975–76 and 1989–90, and degrees conferred in engineering rose approximately 77 percent. However, the largest increase in the *number* of degrees awarded in science was in engineering (an increase of nearly 36,000) followed by computer sciences (an increase of almost 22,000). Other disciplines with larger than average percentage increases in degrees conferred in this period were law, communications, business and management, liberal/general studies, protective services, and area and ethnic studies.

The number of bachelor's degrees in agricultural sciences, life sciences, and physical sciences declined between 1975–76 and 1989–90. Bachelor's degrees in mathematics were volatile, declining 31 percent between 1975–76 and 1980–81, increasing by 47 percent between 1980–81 and 1985–86, and then dropping 10 percent between 1985–86 and 1989–90. Degree production in the fields of education, foreign languages, philosophy and religion, home economics, social sciences, and visual and performing arts declined more than 5 percent.

---

[10] National Science Board, p.48.

[11] National Center for Education Statistics. *Digest of Education Statistics, 1992.* (Washington, D.C.: U.S. Government Printing Office, 1992) p. 209. Data for mathematics, health sciences, agricultural sciences, and computer sciences were either incomplete or unavailable. Also, a high proportion of students listed their major as "unknown."

[12] *Digest, 1992,* p. 210.

[13] *Digest, 1992,* p. 209.

Although the number of bachelor's degrees awarded in science was higher in 1989–90 than in 1975–76, there were significant declines in the last half of the 1980s. Bachelor's degrees conferred in science declined by 16 percent between 1985–86 and 1989–90, compared to a 6 percent increase for bachelor's degrees conferred in all fields. The declines were steepest in computer sciences (35 percent), physical sciences (26 percent), and agricultural sciences (22 percent). The increase in bachelor's degrees was driven by increases in such areas as education, psychology, and social sciences, which increased by 20 percent or more. As a result of these trends, the share of all bachelor's degrees that were awarded in science and mathematics increased from 23 percent in 1975–76 to 30 percent in 1985–86, and then decreased to 24 percent in 1989–90.

Master's Degrees

Between 1975–76 and 1989–90, the number of all master's degrees conferred rose only 4 percent, compared to 13 percent for bachelor's degrees. However, master's degrees in science rose close to 50 percent, compared with a 17 percent increase in bachelor's degrees in science over the same period (table 2–3).

Trends in master's degrees categorized by field somewhat resembled the trends in bachelor's degrees, as the same fields tended to have expanding or decreasing numbers of degrees conferred. For example, master's degrees in computer sciences witnessed the greatest percentage growth during that period, 270 percent. Between 1975–76 and 1989–90, science degrees increased rapidly for health sciences (71 percent) and engineering (52 percent). Master's degrees in agricultural sciences increased, and master's degrees in life sciences, mathematics, and physical sciences decreased.

The most rapidly declining areas of master's degrees between 1975–76 and 1989–90 were non-science and mathematics fields such as foreign languages, education, and social sciences. In contrast, during the same period, non-science and mathematics fields such as communications, law (master's in legal studies), and business and management had increases.

In more recent years, growth in master's degrees in science has slowed. Master's degrees in all fields rose by 12 percent between 1985–86 and 1989–90, by 9 percent in science and 16 percent in mathematics. The increases in science and mathematics master's in the late 1980s contrasted with the significant declines in the number of science and mathematics degrees at the bachelor's level during that same period.

Doctor's Degrees

The overall increase in the number of doctor's degrees awarded between 1975–76 and 1989–90 was 12 percent, similar to that of bachelor's degrees and higher than the increase in master's degrees. As with master's degrees, doctor's degrees in science increased more between 1975–76 and 1989–90 than doctor's degrees in all fields. The number of doctor's degrees conferred in mathematics rose 7 percent, below the national average for all doctor's degrees.

The number of doctor's degrees increased, to a varying extent, in each of the sciences. Between 1975–76 and 1989–90 the largest increases were in health sciences (167 percent), computer sciences (155 percent), and engineering (76 percent). These changes reflect similar large increases in master's degrees in these science fields during the same period.

Doctor's degrees in fields other than science and mathematics declined in this period (table 2–4). For example, doctor's degrees conferred in foreign languages, area and ethnic studies, social sciences, philosophy and religion, liberal/general studies and education decreased by more than 10 percent between 1975–76 and 1989–90.

In contrast to patterns at the lower degree levels, doctor's degrees in science increased at a faster rate than non-science and mathematics fields in more recent years. While doctor's degrees in all fields rose almost 14 percent between 1985–86 and 1989–90, doctor's degrees rose approximately 26 percent in science, and over 23 percent in mathematics. The proportion of doctor's degrees going to science and mathematics graduates also rose steadily, from 36 percent in 1975–76 to 45 percent in 1989–90.

Overall Degree Production

The decreasing numbers of bachelor's degrees awarded in science in the late 1980s may have an effect on future graduate degree production in the 1990s. The increases in master's and doctor's degrees from 1985–86 to 1989–90 may reflect the increasing production up to 1985–86, when bachelor's degrees in science and mathematics rose to a peak of almost 300,000. The decline in bachelor's degrees conferred in science between 1985–86 and 1989–90 will affect the current supply of graduates for the workforce, and may also impact on the future supply of advanced degrees in science.

Characteristics of Science and Mathematics Degree Recipients

The changing characteristics of individuals *enrolled* in higher education (more older, female, and part-time students) were mentioned earlier in this chapter. It is also important to look at the changing characteristics of those *attaining* higher education degrees. Study of science and mathematics in the U.S. has historically been dominated by white men. However, for the academic year ending in 1990, Asian and Hispanic men were as likely or more likely than white men to achieve bachelor's degrees in life sciences, physical sciences, computer sciences and engineering. Black men, and white, Hispanic, and black women were less likely to attain bachelor's degrees in these areas than white men. Asian women were more likely to attain a bachelor's degree in life sciences and physical sciences, and less likely to attain a bachelor's degree in computer sciences and engineering, than white men.[14]

A 1990 NCES report on science majors found that even young women who had worked hard in science and mathematics in high school and had high test scores in these subjects were still less likely to major in science or mathematics in college: "...graduates in these [science and mathematics] fields do not apparently differ much from other graduates in their demographic characteristics, with one important exception: More males than females graduate with such majors...In other words, these were the hardest working, and perhaps the ablest, high school seniors (as measured by high school grades and scores on the High School and Beyond achievement test). They also took more advanced science and mathematics classes in high school and intended to study science, engineering, or mathematics in college more often...differences between graduates in science, engineering, or mathematics and other graduates do not account for the disparities between males and females, however. Regardless of grades in high school or courses taken, females consistently graduate less often than males with a major in one of these fields."[15]

[14] *Condition, 1992*, pp. 70–71.

Summary

Evidence regarding trends in the supply of new graduates with degrees in science and mathematics is mixed. There are some positive signs, such as increases in course taking in science and mathematics "gatekeeping" courses at the secondary level. This pattern suggests that high school students may be entering college better prepared for college level science and mathematics courses than in the past.

Other indicators do not suggest an expanded pipeline. For example, students in middle and high schools still seem to lose interest in science and mathematics as they get older; increased science and mathematics requirements and course taking has not improved student proficiency of 17–year-olds compared to the early 1970s (yet there has been some improvement since the early 1980s); part-time and female students make up a large part of the increase in enrollment in higher education and are less likely to be science and mathematics majors. As a result of these and other factors, numbers of bachelor's degrees conferred in science and mathematics have declined in recent years.

[15] National Center for Education Statistics, *Who Majors in Science?* (Washington, D.C.: U.S. Government Printing Office, 1990) p. 10.

Table 2-1.—National trends in associate degrees, by field: 1982-83 to 1989-90

Field of study	1982-83	1985-86	1989-90	Percent change 1982-83 to 1989-90	Percent change 1982-83 to 1985-86	Percent change 1985-86 to 1989-90
All fields	449,836	446,047	454,679	1.1	-0.8	1.9
Science and mathematics	148,554	150,050	134,624	-9.4	1.0	-10.3
Science	147,777	149,448	133,864	-9.4	1.1	-10.4
Agricultural sciences	7,645	5,741	4,832	-36.8	-24.9	-15.8
Computer sciences	9,670	10,704	7,604	-21.4	10.7	-29.0
Engineering	60,629	63,339	54,131	-10.7	4.5	-14.5
Health sciences	65,749	66,559	64,128	-2.5	1.2	-3.7
Life sciences	981	998	1,034	5.4	1.7	3.6
Physical sciences	3,103	2,107	2,135	-31.2	-32.1	1.3
Mathematics	777	602	760	-2.2	-22.5	26.2
Non-science and mathematics fields	301,282	295,997	320,055	6.2	-1.8	8.1
Area and ethnic studies	23	33	68	195.7	43.5	106.1
Business and management	120,236	117,358	106,980	-11.0	-2.4	-8.8
Communications	3,870	3,984	3,672	-5.1	2.9	-7.8
Education	7,653	7,391	8,018	4.8	-3.4	8.5
Foreign languages	355	437	329	-7.3	23.1	-24.7
Home economics	9,369	9,469	10,230	9.2	1.1	8.0
Law	1,742	2,259	4,547	161.0	29.7	101.3
Letters	638	508	567	-11.1	-20.4	11.6
Liberal/general studies	109,619	107,672	128,721	17.4	-1.8	19.5
Philosophy and religion	193	114	93	-51.8	-40.9	-18.4
Protective services	13,163	12,096	12,848	-2.4	-8.1	6.2
Psychology	1,031	939	1,110	7.7	-8.9	18.2
Public affairs	4,344	3,649	5,228	20.3	-16.0	43.3
Social sciences	2,958	2,540	2,870	-3.0	-14.1	13.0
Visual and performing arts	15,284	13,961	13,923	-8.9	-8.7	-0.3
Other fields, and unclassified	10,804	13,587	20,851	93.0	25.8	53.5
Science and mathematics as a percent of all fields	33.0	33.6	29.6			

Source: National Center for Education Statistics, *Digest of Education Statistics*, 1992, p. 243; *Digest of Education Statistics*, 1989, p. 223; Tables 3-1, 5-1 through 5-6 in this publication; *Historical Trends: State Education Facts, 1969* to 1989, p. 133.

Table 2-2.— National trends in bachelor's degrees, by field: 1975-76 to 1989-90

Field of study	1975-76	1980-81	1985-86	1989-90	Percent change 1975-76 to 1989-90	Percent change 1975-76 to 1980-81	Percent change 1980-81 to 1985-86	Percent change 1985-86 to 1989-90
All fields	925,746	935,140	987,823	1,049,657	13.4	1.0	5.6	6.3
Science and mathematics	216,922	253,601	295,761	249,328	14.9	16.9	16.6	-15.7
Science	200,938	242,523	279,455	234,731	16.8	20.7	15.2	-16.0
Agricultural sciences	19,402	21,886	16,823	13,070	-32.6	12.8	-23.1	-22.3
Computer sciences	5,652	15,121	41,889	27,434	385.4	167.5	177.0	-34.5
Engineering	46,331	75,000	95,953	82,110	77.2	61.9	27.9	-14.4
Health sciences	53,813	63,348	64,535	58,816	9.3	17.7	1.9	-8.9
Life sciences	54,275	43,216	38,524	37,170	-31.5	-20.4	-10.9	-3.5
Physical sciences	21,465	23,952	21,731	16,131	-24.8	11.6	-9.3	-25.8
Mathematics	15,984	11,078	16,306	14,597	-8.7	-30.7	47.2	-10.5
Non-science and mathematics fields	708,824	681,539	692,062	800,329	12.9	-3.8	1.5	15.6
Area and ethnic studies	3,577	2,887	3,060	4,399	23.0	-19.3	6.0	43.8
Business and management	142,379	199,338	238,160	249,081	74.9	40.0	19.5	4.6
Communications	21,282	31,282	43,091	51,283	141.0	47.0	37.8	19.0
Education	154,807	108,309	87,221	104,715	-32.4	-30.0	-19.5	20.1
Foreign languages	15,471	10,319	10,102	11,326	-26.8	-33.3	-2.1	12.1
Home economics	17,409	18,370	15,288	14,987	-13.9	5.5	-16.8	-2.0
Law	531	776	1,197	1,582	197.9	46.1	54.3	32.2
Letters	43,019	33,208	35,434	48,075	11.8	-22.8	6.7	35.7
Liberal/general studies	14,736	18,596	19,248	24,956	69.4	26.2	3.5	29.7
Philosophy and religion	8,447	6,776	6,239	6,848	-18.9	-19.8	-7.9	9.8
Protective services	12,507	13,707	12,704	15,387	23.0	9.6	-7.3	21.1
Psychology	49,908	40,833	40,521	53,586	7.4	-18.2	-0.8	32.2
Public affairs	16,751	18,714	13,878	16,241	-3.0	11.7	-25.8	17.0
Social sciences	126,287	100,345	93,703	116,925	-7.4	-20.5	-6.6	24.8
Visual and performing arts	42,138	40,479	36,949	39,695	-5.8	-3.9	-8.7	7.4
Other fields, and unclassified	39,575	37,600	35,267	41,243	4.2	-5.0	-6.2	16.9
Science and mathematics as a percent of all fields	23.4	27.1	29.9	23.8				

Source: National Center for Education Statistics, *Digest of Education Statistics*, 1992, p. 262.

Table 2-3.— National trends in master's degrees, by field: 1975-76 to 1989-90

Field of study	1975-76	1980-81	1985-86	1989-90	Percent change 1975-76 to 1989-90	Percent change 1975-76 to 1980-81	Percent change 1980-81 to 1985-86	Percent change 1985-86 to 1989-90
All fields	311,771	295,739	288,567	323,844	3.9	-5.1	-2.4	12.2
Science and mathematics	50,075	54,763	66,230	72,203	44.2	9.4	20.9	9.0
Science	46,218	52,196	63,071	68,526	48.3	12.9	20.8	8.6
Agricultural sciences	3,340	4,003	3,801	3,373	1.0	19.9	-5.0	-11.3
Computer sciences	2,603	4,218	8,070	9,643	270.5	62.0	91.3	19.5
Engineering	16,342	16,709	21,661	24,848	52.0	2.2	29.6	14.7
Health sciences	11,885	16,004	18,624	20,354	71.3	34.7	16.4	9.3
Life sciences	6,582	5,978	5,013	4,861	-26.1	-9.2	-16.1	-3.0
Physical sciences	5,466	5,284	5,902	5,447	-0.3	-3.3	11.7	-7.7
Mathematics	3,857	2,567	3,159	3,677	-4.7	-33.4	23.1	16.4
Non-science and mathematics fields	261,696	240,976	222,337	251,641	-3.8	-7.9	-7.7	13.2
Area and ethnic studies	995	804	927	1,198	20.4	-19.2	15.3	29.2
Business and management	42,512	57,898	67,137	77,203	81.6	36.2	16.0	15.0
Communications	3,126	3,105	3,823	4,369	39.8	-0.7	23.1	14.3
Education	128,417	98,938	76,353	86,057	-33.0	-23.0	-22.8	12.7
Foreign languages	3,531	2,104	1,721	1,995	-43.5	-40.4	-18.2	15.9
Home economics	2,179	2,570	2,298	2,153	-1.2	17.9	-10.6	-6.3
Law[1]	1,442	1,832	1,924	1,869	29.6	27.0	5.0	-2.9
Letters	9,468	6,515	6,291	7,223	-23.7	-31.2	-3.4	14.8
Liberal/general studies	1,758	1,085	1,154	1,594	-9.3	-38.3	6.4	38.1
Philosophy and religion	1,356	1,229	1,163	1,326	-2.2	-9.4	-5.4	14.0
Protective services	1,197	1,538	1,074	1,151	-3.8	28.5	-30.2	7.2
Psychology	7,811	7,998	8,293	9,231	18.2	2.4	3.7	11.3
Public affairs	16,117	18,524	16,300	17,993	11.6	14.9	-12.0	10.4
Social sciences	15,824	11,855	10,428	11,419	-27.8	-25.1	-12.0	9.5
Visual and performing arts	8,817	8,629	8,416	8,546	-3.1	-2.1	-2.5	1.5
Other fields, and unclassified	17,146	16,352	15,035	18,314	6.8	-4.6	-8.1	21.8
Science and mathematics as a percent of all fields	16.1	18.5	23.0	22.3				

[1] A master's degree in law is a degree in legal studies. A degree to practice law, such as a J.D., would be a first-professional degree.

Source: National Center for Education Statistics, *Digest of Education Statistics, 1992,* p. 263.

Table 2-4.— National trends in doctor's degrees, by field: 1975-76 to 1989-90

Field of study	1975-76	1980-81	1985-86	1989-90	Percent change 1975-76 to 1989-90	Percent change 1975-76 to 1980-81	Percent change 1980-81 to 1985-86	Percent change 1985-86 to 1989-90
All fields	34,064	32,958	33,653	38,238	12.3	-3.2	2.1	13.6
Science and mathematics	12,249	12,294	13,804	17,330	41.5	0.4	12.3	25.5
Science	11,393	11,566	13,062	16,415	44.1	1.5	12.9	25.7
Agricultural sciences	928	1,067	1,158	1,272	37.1	15.0	8.5	9.8
Computer sciences	244	252	344	623	155.3	3.3	36.5	81.1
Engineering	2,821	2,561	3,410	4,965	76.0	-9.2	33.2	45.6
Health sciences	577	827	1,241	1,543	167.4	43.3	50.1	24.3
Life sciences	3,392	3,718	3,358	3,844	13.3	9.6	-9.7	14.5
Physical sciences	3,431	3,141	3,551	4,168	21.5	-8.5	13.1	17.4
Mathematics	856	728	742	915	6.9	-15.0	1.9	23.3
Non-science and mathematics fields	21,815	20,664	19,849	20,908	-4.2	-5.3	-3.9	5.3
Area and ethnic studies	188	162	157	128	-31.9	-13.8	-3.1	-18.5
Business and management	953	842	969	1,142	19.8	-11.6	15.1	17.9
Communications	204	182	223	269	31.9	-10.8	22.5	20.6
Education	7,778	7,900	7,110	6,922	-11.0	1.6	-10.0	-2.6
Foreign languages	864	588	448	512	-40.7	-31.9	-23.8	14.3
Home economics	178	247	311	303	70.2	38.8	25.9	-2.6
Law	76	60	54	113	48.7	-21.1	-10.0	109.3
Letters	1,884	1,380	1,215	1,266	-32.8	-26.8	-12.0	4.2
Liberal/general studies	36	23	38	31	-13.9	-36.1	65.2	-18.4
Philosophy and religion	554	410	477	432	-22.0	-26.0	16.3	-9.4
Protective services	9	21	21	37	311.1	133.3	0.0	76.2
Psychology	2,581	2,955	3,088	3,353	29.9	14.5	4.5	8.6
Public affairs	298	388	385	495	66.1	30.2	-0.8	28.6
Social sciences	4,154	3,114	2,955	3,023	-27.2	-25.0	-5.1	2.3
Visual and performing arts	620	654	722	842	35.8	5.5	10.4	16.6
Other fields, and unclassified	1,438	1,738	1,676	2,040	41.9	20.9	-3.6	21.7
Science and mathematics as a percent of all fields	36.0	37.3	41.0	45.3				

Source: National Center for Education Statistics, *Digest of Education Statistics*, 1992, p. 264.

3. Supply of Graduates in Mathematics: Associate, Bachelor's, Master's, and Doctor's Degrees by Region and State

A look at trends in the awarding of mathematics degrees, and the various growth rates at the regional and state level, provides an indication of the supply of individuals with mathematical training. Regionally, the Midwest and the West both had increases in bachelor's and master's degrees awarded in mathematics between 1975–76 and 1989–90, while the Northeast and the South both declined in numbers of degrees conferred in mathematics at those levels.

Associate Degrees

Text table 1 shows that the number of associate degrees conferred in mathematics between 1982–83[1] and 1989–90 was small and unevenly spread across the four regions. The West accounted for over half of all associate degrees in mathematics. The number of associates awarded in mathematics rose during the late 1980s despite losses in the Northeast, Midwest, and South, because of an increase of 47 percent in the West. (Table 3–1 presents data on associate degrees in mathematics by state.)

[1] The academic year 1982–83 is the first year for which the field of study data are consistent for associate degrees.

Text table 1.—Mathematics degrees by region and degree level: 1975-76 to 1989-90

Degree level and region	1975-76	1982-83	1985-86	1989-90	Percent change: 1982-83 to 1989-90	Percent change: 1982-83 to 1985-86	Percent change: 1985-86 to 1989-90
Associate degrees							
United States	—	777	602	760	-2.2	-22.5	26.2
Northeast	—	28	45	41	46.4	60.7	-8.9
Midwest	—	134	29	26	-80.6	-78.4	-10.3
South	—	91	164	159	74.7	80.2	-3.0
West	—	524	364	534	1.9	-30.5	46.7
U.S.S.S.	—	0	0	—	(1)	(1)	(1)

Degree level and region	1975-76	1980-81	1985-86	1989-90	Percent change: 1975-76 to 1989-90	Percent change: 1975-76 to 1980-81	Percent change: 1980-81 to 1985-86	Percent change: 1985-86 to 1989-90
Bachelor's degrees								
United States	15,984	11,078	16,306	14,597	-8.7	-30.7	47.2	-10.5
Northeast	5,193	3,315	4,566	3,603	-30.6	-36.2	37.7	-21.1
Midwest	3,946	2,764	4,110	4,116	4.3	-30.0	48.7	0.1
South	4,316	3,191	4,330	3,978	-7.8	-26.1	35.7	-8.1
West	2,418	1,712	3,164	2,806	16.0	-29.2	84.8	-11.3
U.S.S.S.	111	96	136	94	-15.3	-13.5	41.7	-30.9
Master's degrees								
United States	3,857	2,567	3,159	3,677	-4.7	-33.4	23.1	16.4
Northeast	1,301	682	888	902	-30.7	-47.6	30.2	1.6
Midwest	933	689	860	1,122	20.3	-26.2	24.8	30.5
South	1,025	767	826	986	-3.8	-25.2	7.7	19.4
West	595	427	583	667	12.1	-28.2	36.5	14.4
U.S.S.S.	3	2	2	0	-100.0	-33.3	0.0	-100.0
Doctor's degrees								
United States	856	728	742	915	6.9	-15.0	1.9	23.3
Northeast	281	233	244	279	-0.7	-17.1	4.7	14.3
Midwest	229	174	183	242	5.7	-24.0	5.2	32.2
South	159	150	162	214	34.6	-5.7	8.0	32.1
West	187	171	153	180	-3.7	-8.6	-10.5	17.6
U.S.S.S.	0	0	0	0	(1)	(1)	(1)	(1)

1 Insufficient data for calculating a percent change.
—Data not available.
Note: U.S.S.S.=U. S. Service Schools.

SOURCE: U.S. Department of Education, National Center for Education Statistics, HEGIS, "Degrees and Other Formal Awards Conferred" surveys; and IPEDS," Completions" surveys.

Bachelor's Degrees

Regions

Recent patterns in awarding of bachelor's degrees in mathematics have mirrored trends in science degrees, with increases in the early 1980s and declining numbers in the late 1980s. Between 1975–76 and 1989–90, bachelor's in mathematics fell in the Northeast (31 percent) and in the South (8 percent), while rising in the West (16 percent) and in the Midwest (4 percent) (text table 1). During the later period between 1985–86 and 1989–90, bachelor's degrees conferred in mathematics dropped in the Northeast (21 percent), West (11 percent), and the South (8 percent), and increased slightly in the Midwest. In 1975–76, the Northeast was the largest producer of bachelor's degrees in mathematics, but it had fallen to third place behind the Midwest and South in 1989–90.

Chart 2 — Bachelor's degrees in mathematics, by region: 1975–76 to 1989–90

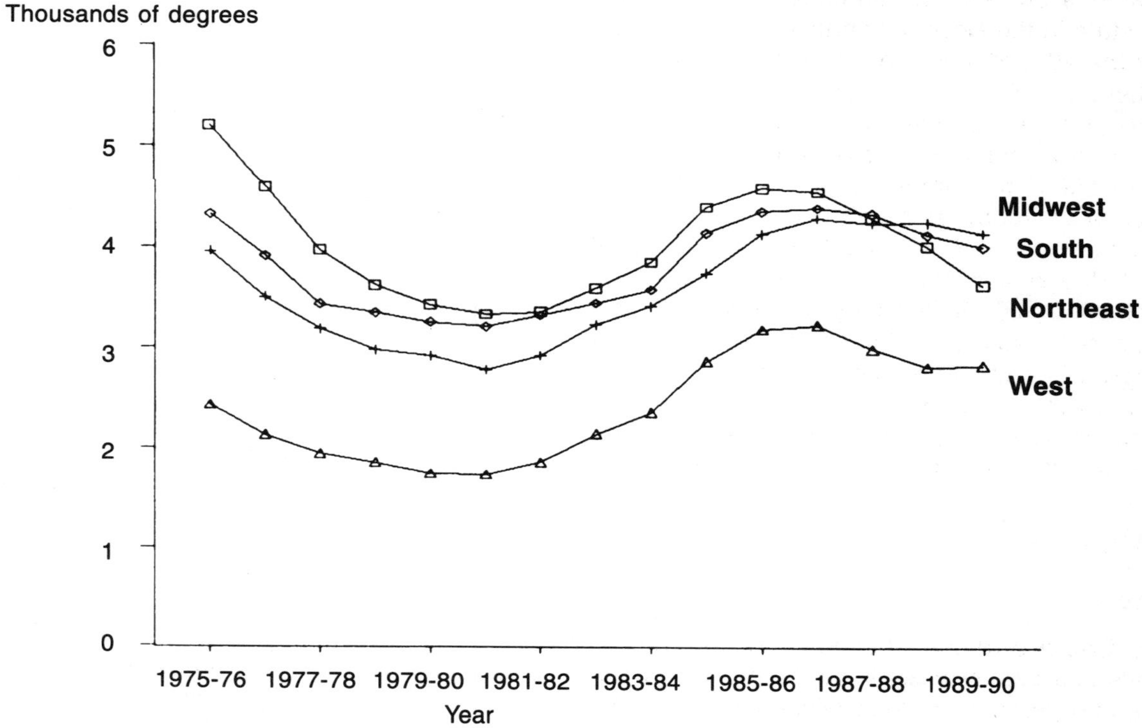

Source: U.S. Department of Education, National Center for Education Statistics, HEGIS, "Degrees and Other Formal Awards Conferred" surveys; and IPEDS, "Completions" surveys.

States

Table 3–2 presents data on bachelor's degrees in mathematics by state. In the Northeast, bachelor's degrees conferred in mathematics between 1975–76 and 1989–90 declined by 30 percent or more in every state except Pennsylvania and Vermont. In contrast to the Northeast, bachelor's degrees in mathematics rose in every state in the West except Arizona, Idaho, and New Mexico.

The early 1980s' increases in bachelor's degrees in mathematics were reflected in 44 states. However, in the late 1980s, bachelor's in mathematics dropped in 28 states, the District of Columbia, and the Service Schools. Every state in the Northeast declined in awarding bachelor's in mathematics, from 2 percent in Vermont to 45 percent in New Hampshire, whereas only two states in the Midwest experienced decreases during this period. In 1989–90, three states, California, New York, and Pennsylvania, each awarded over 1,000 bachelor's degrees in mathematics and accounted for 25 percent of such degrees.

Master's Degrees

Regions

There was great variation among the regions in the trends between 1975–76 and 1989–90. Text table 1 shows that master's in mathematics dropped by 31 percent in the Northeast, while there was a relatively small decline in the South (4 percent), and increases in the Midwest (20 percent) and the West (12 percent). Master's degrees in mathematics declined in every region in the country in the late 1970s, but the decline was much steeper in the Northeast than in the other three regions. However, master's degrees conferred in mathematics increased in every region between 1980–81 and 1985–86 and between 1985–86 and 1989–90.

States

Table 3–3 shows that between 1975–76 and 1980–81 master's degrees in mathematics decreased in every state in the Northeast and Midwest, and in 18 out of the 29 states in the South and West, and in the District of Columbia. During the early 1980s, however, master's degrees in mathematics increased in almost two-thirds of the states and the District of Columbia, and the numbers of master's conferred in mathematics increased in nearly three-fourths of the states between 1985–86 and 1989–90. In 1989–90 New York awarded the most master's degrees in mathematics (430), followed by California (324), Ohio (248), and Illinois (235). These four states accounted for one-third of all the master's degrees awarded in mathematics in that year. However, the number awarded in New York fell by over one-third between 1975–76 and 1989–90.

Doctor's Degrees

Regions

The South and the Midwest had increases in doctor's degrees awarded in mathematics of 35 percent and 6 percent, respectively, between 1975–76 and 1989–90. The number of doctor's degrees in mathematics in the Northeast declined 1 percent and in the West the number declined 4 percent during this period (text table 1). Between 1985–86 and 1989–90 doctor's degrees conferred in mathematics rose in every region in the country from 14 percent in the Northeast, to 18 percent in the West, to 32 percent in the Midwest and South. The Northeast awarded the greatest number of doctor's degrees in mathematics throughout the period. The South, however, moved past the West into third place as a result of its substantial growth in such degrees.

States

Table 3–4 shows that 19 states and the Service Schools awarded 5 or less doctor's degrees in mathematics (or provided no data) in 1989–90. This was similar to 1975–76. New York (120 degrees) and California (101 degrees) accounted for 24 percent of the total doctor's degrees awarded in mathematics in 1989–90.

Summary

In 1989–90, the total number of bachelor's degrees conferred in mathematics in the Nation was almost 15,000, a small percentage of the over 1 million bachelor's degrees conferred. However, since mathematics is important, not only as a subject itself but also in the understanding of science and technology, trends in degrees awarded help to describe the future supply of scientific and technically trained personnel.

Mathematics is one subject in which regional changes in trends of degrees conferred are clearly apparent. The West awarded the fewest bachelor's, master's, and doctor's degrees in this subject. The Northeast, once at the front in the awarding of degrees in mathematics above the associate level has slipped to third in awarding bachelor's and master's, with the numbers of degrees awarded in mathematics dropping. In 1989–90 doctor's degrees in mathematics in the Northeast were at the 1975–76 level, but this region still awards the largest number of doctor's degrees in mathematics.

While the national trend in the awarding of doctor's degrees in mathematics is upward, the small numbers of doctor's degrees awarded in mathematics in almost half the states could be an area of concern. Future research could examine the reasons for these regional changes and state-by-state declines at all levels, and the effects of these changes on the mathematics profession.

Table 3-1.—Associate degrees conferred in mathematics, by region and state: 1982-83 to 1989-90

Region and state	1975-76	1976-77	1977-78	1978-79	1979-80	1980-81	1981-82	1982-83	1983-84	1984-85	1985-86	1986-87	1987-88	1988-89	1989-90	Percent change 1982-83 to 1989-90	Percent change 1982-83 to 1985-86	Percent change 1985-86 to 1989-90
United States	—	—	—	—	—	—	—	777	783	789	602	667	684	654	760	-2.2	-22.5	26.2
Northeast	—	—	—	—	—	—	—	28	36	50	45	49	66	38	41	46.4	60.7	-8.9
Connecticut	—	—	—	—	—	—	—	0	0	0	0	—	—	—	—	(1)	(1)	(1)
Maine	—	—	—	—	—	—	—	0	0	0	0	—	—	—	—	(1)	(1)	(1)
Massachusetts	—	—	—	—	—	—	—	6	4	4	14	9	15	8	7	16.7	133.3	-50.0
New Hampshire	—	—	—	—	—	—	—	0	0	0	0	—	—	—	—	(1)	(1)	(1)
New Jersey	—	—	—	—	—	—	—	1	1	4	2	5	1	3	1	0.0	100.0	-50.0
New York	—	—	—	—	—	—	—	10	15	11	11	9	17	9	9	-10.0	10.0	-18.2
Pennsylvania	—	—	—	—	—	—	—	11	16	31	18	26	33	18	24	118.2	63.6	33.3
Rhode Island	—	—	—	—	—	—	—	0	0	0	0	—	—	—	—	(1)	(1)	(1)
Vermont	—	—	—	—	—	—	—	0	0	0	0	—	—	—	—	(1)	(1)	(1)
Midwest	—	—	—	—	—	—	—	134	114	65	29	42	47	25	26	-80.6	-78.4	-10.3
Illinois	—	—	—	—	—	—	—	52	47	28	0	1	—	—	—	(1)	-100.0	(1)
Indiana	—	—	—	—	—	—	—	1	3	4	2	3	4	—	—	(1)	100.0	(1)
Iowa	—	—	—	—	—	—	—	0	4	0	0	—	—	—	—	(1)	(1)	(1)
Kansas	—	—	—	—	—	—	—	23	17	5	9	13	4	5	11	-52.2	-60.9	22.2
Michigan	—	—	—	—	—	—	—	25	11	19	13	20	26	6	8	-68.0	-48.0	-38.5
Minnesota	—	—	—	—	—	—	—	2	1	4	0	0	—	—	—	(1)	-100.0	(1)
Missouri	—	—	—	—	—	—	—	0	8	0	0	—	—	0	2	(1)	(1)	(1)
Nebraska	—	—	—	—	—	—	—	1	2	1	1	—	—	5	—	(1)	0.0	(1)
North Dakota	—	—	—	—	—	—	—	5	2	2	1	2	1	—	—	(1)	-80.0	(1)
Ohio	—	—	—	—	—	—	—	17	11	2	3	3	12	9	5	-70.6	-82.4	66.7
South Dakota	—	—	—	—	—	—	—	8	8	0	0	—	—	—	—	(1)	-100.0	(1)
Wisconsin	—	—	—	—	—	—	—	0	0	0	0	—	—	—	—	(1)	(1)	(1)
South	—	—	—	—	—	—	—	91	144	180	164	156	146	165	159	74.7	80.2	-3.0
Alabama	—	—	—	—	—	—	—	12	17	19	33	34	35	30	22	83.3	175.0	-33.3
Arkansas	—	—	—	—	—	—	—	5	0	0	0	—	—	—	—	(1)	-100.0	(1)
Delaware	—	—	—	—	—	—	—	0	1	0	2	—	—	—	1	(1)	(1)	-50.0
District of Columbia	—	—	—	—	—	—	—	0	0	0	0	—	—	—	—	(1)	(1)	(1)
Florida	—	—	—	—	—	—	—	0	0	0	0	—	5	—	—	(1)	(1)	(1)
Georgia	—	—	—	—	—	—	—	0	3	2	6	—	—	—	1	(1)	(1)	-83.3
Kentucky	—	—	—	—	—	—	—	1	3	20	20	15	5	7	8	700.0	1900.0	-60.0
Louisiana	—	—	—	—	—	—	—	0	0	0	1	1	2	3	3	(1)	(1)	200.0
Maryland	—	—	—	—	—	—	—	0	0	0	0	—	—	—	—	(1)	(1)	(1)
Mississippi	—	—	—	—	—	—	—	7	8	6	3	4	2	2	4	-42.9	-57.1	33.3
North Carolina	—	—	—	—	—	—	—	7	0	5	0	1	0	—	—	(1)	-100.0	(1)
Oklahoma	—	—	—	—	—	—	—	0	43	49	39	43	33	44	42	(1)	(1)	7.7
South Carolina	—	—	—	—	—	—	—	8	0	0	0	—	—	—	—	(1)	-100.0	(1)
Tennessee	—	—	—	—	—	—	—	5	1	0	0	1	1	0	—	(1)	-100.0	(1)
Texas	—	—	—	—	—	—	—	44	65	74	60	57	63	79	78	77.3	36.4	30.0
Virginia	—	—	—	—	—	—	—	2	3	3	0	—	—	—	—	(1)	-100.0	(1)
West Virginia	—	—	—	—	—	—	—	0	0	2	0	—	—	—	—	(1)	(1)	(1)
West	—	—	—	—	—	—	—	524	489	494	364	420	425	426	534	1.9	-30.5	46.7
Alaska	—	—	—	—	—	—	—	10	5	1	0	—	—	—	—	(1)	-100.0	(1)
Arizona	—	—	—	—	—	—	—	5	5	6	3	5	6	1	1	-80.0	-40.0	-66.7
California	—	—	—	—	—	—	—	445	414	416	297	388	393	389	491	10.3	-33.3	65.3
Colorado	—	—	—	—	—	—	—	39	27	25	25	—	—	—	—	(1)	-35.9	(1)
Hawaii	—	—	—	—	—	—	—	0	0	0	0	—	—	—	—	(1)	(1)	(1)

Table 3-1.—Continued

Region and state	1975-76	1976-77	1977-78	1978-79	1979-80	1980-81	1981-82	1982-83	1983-84	1984-85	1985-86	1986-87	1987-88	1988-89	1989-90	Percent change: 1982-83 to 1989-90	Percent change: 1982-83 to 1985-86	Percent change: 1985-86 to 1989-90
West Continued																		
Idaho	—	—	—	—	—	—	—	7	9	11	13	11	7	14	15	114.3	85.7	15.4
Montana	—	—	—	—	—	—	—	2	0	2	1	4	5	2	8	300.0	-50.0	700.0
Nevada	—	—	—	—	—	—	—	0	0	0	0	—	—	—	—	(1)	(1)	(1)
New Mexico	—	—	—	—	—	—	—	1	1	2	1	—	—	—	—	(1)	0.0	(1)
Oregon	—	—	—	—	—	—	—	5	6	8	5	3	5	6	4	-20.0	0.0	-20.0
Utah	—	—	—	—	—	—	—	0	0	0	0	—	2	3	8	(1)	(1)	(1)
Washington	—	—	—	—	—	—	—	2	7	7	2	—	—	—	—	(1)	0.0	(1)
Wyoming	—	—	—	—	—	—	—	8	15	16	17	9	7	11	7	-12.5	112.5	-58.8
U.S. Service Schools	—	—	—	—	—	—	—	0	0	0	0	0	—	—	—	(1)	(1)	(1)
Outlying Areas	—	—	—	—	—	—	—	0	0	0	3	0	0	0	0	(1)	(1)	-100.0
American Samoa	—	—	—	—	—	—	—	—	—	—	—	—	—	—	—	(1)	(1)	(1)
Guam	—	—	—	—	—	—	—	0	0	0	0	—	—	—	—	(1)	(1)	(1)
Northern Marianas	—	—	—	—	—	—	—	—	—	—	—	—	—	—	—	(1)	(1)	(1)
Puerto Rico	—	—	—	—	—	—	—	0	0	0	3	—	—	—	—	(1)	(1)	(1)
Trust Territories	—	—	—	—	—	—	—	—	—	—	—	—	—	—	—	(1)	(1)	(1)
Virgin Islands	—	—	—	—	—	—	—	0	0	0	0	—	—	—	—	(1)	(1)	(1)

[1]Insufficient data for calculating a percent change.

—Data not available.

SOURCE: U.S. Department of Education, National Center for Education Statistics, HEGIS, "Degrees and Other Formal Awards Conferred" surveys; and IPEDS,"Completions" surveys.

Table 3-2.— Bachelor's degrees conferred in mathematics, by region and state: 1975-76 to 1989-90

Region and state	1975-76	1976-77	1977-78	1978-79	1979-80	1980-81	1981-82	1982-83	1983-84	1984-85	1985-86	1986-87	1987-88	1988-89	1989-90	Percent change 1975-76 to 1989-90	1975-76 to 1980-81	1980-81 to 1985-86	1985-86 to 1989-90
United States . . .	15,984	14,196	12,569	11,806	11,378	11,078	11,599	12,453	13,211	15,146	16,306	16,444	15,904	15,218	14,597	-8.7	-30.7	47.2	-10.5
Northeast	5,193	4,587	3,954	3,609	3,409	3,315	3,342	3,575	3,833	4,376	4,566	4,528	4,266	3,986	3,603	-30.6	-36.2	37.7	-21.1
Connecticut . .	269	250	220	176	160	170	190	181	254	303	307	261	233	194	180	-33.1	-36.8	80.6	-41.4
Maine	149	98	98	83	107	84	90	66	94	104	83	74	80	77	76	-49.0	-43.6	-1.2	-8.4
Massachusetts	831	798	679	681	623	631	667	663	705	694	664	709	675	608	569	-31.5	-24.1	5.2	-14.3
New Hampshire	133	139	111	107	86	88	77	101	104	110	139	137	123	107	77	-42.1	-33.8	58.0	-44.6
New Jersey . . .	519	441	366	345	306	289	282	280	267	273	336	289	265	270	264	-49.1	-44.3	16.3	-21.4
New York	1,875	1,641	1,489	1,303	1,184	1,164	1,123	1,286	1,339	1,663	1,710	1,613	1,435	1,380	1,206	-35.7	-37.9	46.9	-29.5
Pennsylvania .	1,210	1,046	802	766	796	703	710	778	821	989	1,115	1,238	1,264	1,164	1,056	-12.7	-41.9	58.6	-5.3
Rhode Island .	134	103	125	107	98	131	143	145	170	133	126	120	105	114	91	-32.1	-2.2	-3.8	-27.8
Vermont	73	71	64	41	49	55	60	75	79	107	86	87	86	72	84	15.1	-24.7	56.4	-2.3
Midwest	3,946	3,490	3,174	2,958	2,902	2,764	2,906	3,215	3,394	3,722	4,110	4,265	4,214	4,223	4,116	4.3	-30.0	48.7	0.1
Illinois	782	731	625	603	575	565	618	701	793	829	941	816	848	750	726	-7.2	-27.7	66.5	-22.8
Indiana	422	376	330	362	359	305	330	386	401	437	456	486	418	363	355	-15.9	-27.7	49.5	-22.1
Iowa	199	171	127	136	164	138	134	177	153	214	248	280	258	301	261	31.2	-30.7	79.7	5.2
Kansas	163	138	128	111	120	92	109	117	122	135	128	163	175	188	149	-8.6	-43.6	39.1	16.4
Michigan	526	473	450	370	323	336	337	390	428	472	516	512	523	602	587	11.6	-36.1	53.6	13.8
Minnesota	298	293	304	253	284	294	282	340	365	352	378	403	402	467	450	51.0	-1.3	28.6	19.0
Missouri	263	234	230	213	199	208	195	175	162	181	205	179	236	216	212	-19.4	-20.9	-1.4	3.4
Nebraska	168	155	117	92	88	94	104	108	98	147	160	168	170	139	167	-0.6	-44.0	70.2	4.4
North Dakota .	34	25	31	21	29	34	42	33	29	43	60	63	62	61	62	82.4	0.0	76.5	3.3
Ohio	642	530	482	465	467	403	444	474	475	480	496	600	542	573	555	-13.6	-37.2	23.1	11.9
South Dakota .	85	45	52	41	48	47	47	47	50	58	80	71	72	60	82	-3.5	-44.7	70.2	2.5
Wisconsin	364	319	298	291	246	248	264	267	318	374	442	524	508	503	510	40.1	-31.9	78.2	15.4
South	4,316	3,896	3,422	3,337	3,240	3,191	3,306	3,427	3,561	4,122	4,330	4,367	4,307	4,101	3,978	-7.8	-26.1	35.7	-8.1
Alabama	228	211	179	197	168	187	163	139	178	211	245	251	260	266	275	20.6	-18.0	31.0	12.2
Arkansas	108	107	71	98	78	77	83	59	54	72	76	87	75	86	78	-27.8	-28.7	-1.3	2.6
Delaware	33	41	32	32	33	28	20	29	25	43	41	40	45	38	29	-12.1	-15.2	46.4	-29.3
District of Columbia . . .	71	80	71	68	56	59	71	54	36	41	43	42	50	34	34	-52.1	-16.9	-27.1	-20.9
Florida	287	276	256	215	230	239	246	237	283	287	247	325	311	265	285	-0.7	-16.7	3.3	15.4
Georgia	256	251	175	247	195	224	240	292	256	298	297	299	297	273	242	-5.5	-12.5	32.6	-18.5
Kentucky	198	107	121	111	102	105	122	140	152	235	270	247	247	235	252	27.3	-47.0	157.1	-6.7
Louisiana	136	140	130	88	93	88	109	93	117	106	128	116	97	120	134	-1.5	-35.3	45.5	4.7
Maryland	283	237	217	219	197	229	233	218	213	248	263	239	239	265	218	-23.0	-19.1	14.8	-17.1
Mississippi . . .	138	126	115	90	89	80	72	71	86	79	121	116	167	137	130	-5.8	-42.0	51.3	7.4
North Carolina	517	479	411	426	404	420	454	525	584	661	619	530	567	487	477	-7.7	-18.8	47.4	-22.9
Oklahoma	226	172	139	136	129	131	132	109	121	141	144	160	133	170	167	-26.1	-42.0	9.9	16.0
South Carolina	268	247	221	238	221	213	212	273	236	290	269	345	246	345	230	-14.2	-20.5	26.3	-14.5
Tennessee . . .	277	287	231	207	238	215	204	197	215	267	294	298	315	297	272	-1.8	-22.4	36.7	-7.5
Texas	764	711	649	587	616	557	523	566	531	605	687	725	709	612	680	-11.0	-27.1	23.3	-1.0
Virginia	417	332	341	322	341	291	364	386	420	458	502	473	486	410	418	0.2	-30.2	72.5	-16.7
West Virginia .	109	92	63	56	50	48	58	39	54	80	84	74	63	61	57	-47.7	-56.0	75.0	-32.1
West	2,418	2,121	1,932	1,838	1,722	1,712	1,840	2,129	2,343	2,846	3,164	3,210	2,973	2,794	2,806	16.0	-29.2	84.8	-11.3
Alaska	7	8	7	8	9	11	10	11	15	14	15	25	16	20	26	271.4	57.1	36.4	73.3
Arizona	91	74	81	79	77	78	59	56	47	51	66	70	80	85	90	-1.1	-14.3	-15.4	36.4
California	1,338	1,185	1,057	1,022	966	957	1,009	1,150	1,295	1,527	1,734	1,771	1,573	1,419	1,387	3.7	-28.5	81.2	-20.0
Colorado	201	174	167	183	150	179	191	219	254	320	327	357	399	366	362	80.1	-10.9	82.7	10.7
Hawaii	51	38	50	25	16	20	37	29	29	45	54	41	41	50	54	5.9	-60.8	170.0	0.0
Idaho	46	40	32	30	25	26	32	37	31	39	77	63	53	48	46	0.0	-43.5	196.2	-40.3

Table 3-2.— Continued

Region and state	1975-76	1976-77	1977-78	1978-79	1979-80	1980-81	1981-82	1982-83	1983-84	1984-85	1985-86	1986-87	1987-88	1988-89	1989-90	Percent change: 1975-76 to 1989-90	Percent change: 1975-76 to 1980-81	Percent change: 1980-81 to 1985-86	Percent change: 1985-86 to 1989-90
West Continued																			
Montana	60	48	41	39	54	38	46	67	50	72	82	98	78	74	83	38.3	-36.7	115.8	1.2
Nevada	12	7	18	11	14	15	10	9	11	21	10	15	14	5	18	50.0	25.0	-33.3	80.0
New Mexico	96	104	66	59	68	46	61	70	60	75	96	69	78	94	80	-16.7	-52.1	108.7	-16.7
Oregon	129	134	99	108	101	86	117	141	154	180	161	172	157	147	157	21.7	-33.3	87.2	-2.5
Utah	110	89	115	91	65	76	73	84	126	164	189	199	179	159	167	51.8	-30.9	148.7	-11.6
Washington	259	212	193	172	168	171	186	247	264	330	327	310	283	311	311	20.1	-34.0	91.2	-4.9
Wyoming	18	8	6	11	9	9	9	9	7	8	26	20	22	16	25	38.9	-50.0	188.9	-3.8
U.S. Service Schools	111	102	87	64	105	96	205	107	80	80	136	74	144	114	94	-15.3	-13.5	41.7	-30.9
Outlying Areas	101	107	132	95	95	95	109	104	131	121	82	71	93	77	77	-23.8	-5.9	-13.7	-6.1
American Samoa	—	—	—	—	—	—	—	—	—	—	—	—	—	—	—	(1)	(1)	(1)	(1)
Guam	5	9	3	1	6	1	2	3	3	6	1	7	9	2	2	-60.0	-80.0	0.0	100.0
Northern Marianas	—	—	—	—	—	—	—	—	—	—	—	—	—	—	—	(1)	(1)	(1)	(1)
Puerto Rico	93	98	126	93	83	92	105	97	122	109	79	58	79	69	69	-25.8	-1.1	-14.1	-12.7
Trust Territories	—	—	—	—	—	—	—	—	—	—	—	—	—	—	—	(1)	(1)	(1)	(1)
Virgin Islands	3	—	3	1	6	2	2	4	6	6	2	6	5	6	6	100.0	-33.3	0.0	200.0

[1]Insufficient data for calculating a percent change.

—Data not available.

SOURCE: U.S. Department of Education, National Center for Education Statistics, HEGIS, "Degrees and Other Formal Awards Conferred" surveys; and IPEDS,"Completions" surveys.

Table 3-3.—Master's degrees conferred in mathematics, by region and state: 1975-76 to 1989-90

Region and state	1975-76	1976-77	1977-78	1978-79	1979-80	1980-81	1981-82	1982-83	1983-84	1984-85	1985-86	1986-87	1987-88	1988-89	1989-90	Percent change 1975-76 to 1989-90	Percent change 1975-76 to 1980-81	Percent change 1980-81 to 1985-86	Percent change 1985-86 to 1989-90
United States	3,857	3,695	3,373	3,036	2,860	2,567	2,727	2,837	2,741	2,882	3,159	3,319	3,442	3,447	3,677	-4.7	-33.4	23.1	16.4
Northeast	1,301	1,135	992	848	688	682	735	786	721	774	888	887	840	872	902	-30.7	-47.6	30.2	1.6
Connecticut	50	61	50	38	32	30	40	38	36	23	41	47	43	47	46	-8.0	-40.0	36.7	12.2
Maine	7	2	4	3	6	5	4	2	3	3	5	3	2	1	4	-42.9	-28.6	0.0	-20.0
Massachusetts	148	135	125	100	88	72	94	115	139	193	224	143	184	144	143	-3.4	-51.4	211.1	-36.2
New Hampshire	25	20	24	16	12	11	14	14	15	10	16	9	17	14	23	-8.0	-56.0	45.5	43.8
New Jersey	131	104	107	81	72	75	70	59	43	45	45	68	50	51	44	-66.4	-42.7	-40.0	-2.2
New York	648	610	478	422	326	349	358	416	335	343	382	424	386	436	430	-33.6	-46.1	9.5	12.6
Pennsylvania	239	170	177	159	117	106	123	114	126	128	129	143	113	145	171	-28.5	-55.6	21.7	32.6
Rhode Island	44	26	16	23	28	31	30	23	18	23	30	39	37	21	31	-29.5	-29.5	-3.2	3.3
Vermont	9	7	11	6	7	3	2	5	6	6	16	11	8	13	10	11.1	-66.7	433.3	-37.5
Midwest	933	1,018	925	840	830	689	697	716	821	881	860	949	1,078	1,074	1,122	20.3	-26.2	24.8	30.5
Illinois	211	257	229	186	163	135	137	186	200	189	213	204	272	207	235	11.4	-36.0	57.8	10.3
Indiana	137	117	136	112	119	105	95	90	94	108	115	111	110	125	108	-21.2	-23.4	9.5	-6.1
Iowa	57	60	43	65	48	54	55	35	63	72	75	74	106	99	104	82.5	-5.3	38.9	38.7
Kansas	36	37	37	27	25	32	32	24	24	34	19	43	39	34	34	-5.6	-11.1	-40.6	78.9
Michigan	191	166	133	134	146	105	109	85	110	112	93	107	128	140	143	-25.1	-45.0	-11.4	53.8
Minnesota	22	31	37	19	33	12	34	40	38	39	42	45	48	55	51	131.8	-45.5	250.0	21.4
Missouri	55	62	58	67	50	48	45	42	29	43	32	38	36	32	33	-40.0	-12.7	-33.3	3.1
Nebraska	20	23	23	19	23	11	10	19	14	25	21	40	40	55	69	245.0	-45.0	90.9	228.6
North Dakota	7	4	6	3	1	6	5	4	4	6	6	7	7	8	17	142.9	-14.3	0.0	183.3
Ohio	115	148	135	133	145	106	120	132	189	196	200	208	213	246	248	115.7	-7.8	88.7	24.0
South Dakota	9	14	5	4	3	5	8	7	7	3	2	7	8	11	9	0.0	-44.4	-60.0	350.0
Wisconsin	73	99	83	71	74	70	47	52	49	54	42	65	71	62	71	-2.7	-4.1	-40.0	69.0
South	1,025	954	916	846	893	767	841	803	682	750	826	931	944	924	986	-3.8	-25.2	7.7	19.4
Alabama	45	39	27	16	27	30	20	11	17	20	19	22	32	39	46	2.2	-33.3	-36.7	142.1
Arkansas	21	25	21	25	12	13	11	14	10	7	19	13	17	12	17	-19.0	-38.1	46.2	-10.5
Delaware	7	8	4	6	7	7	4	5	2	9	5	3	3	10	10	42.9	0.0	-28.6	100.0
District of Columbia	45	43	47	75	57	33	109	76	50	54	65	64	54	8	17	-62.2	-26.7	97.0	-73.8
Florida	94	85	67	67	87	63	72	88	66	61	34	73	82	74	109	16.0	-33.0	-46.0	220.6
Georgia	76	76	70	51	82	47	47	55	47	39	59	74	75	73	88	15.8	-38.2	25.5	49.2
Kentucky	47	31	44	41	41	32	26	23	19	28	28	33	32	30	33	-29.8	-31.9	-12.5	17.9
Louisiana	38	33	28	32	33	44	51	31	41	46	48	51	48	57	69	81.6	15.8	9.1	43.8
Maryland	56	56	59	58	45	45	39	41	37	42	54	52	54	52	56	0.0	-19.6	20.0	3.7
Mississippi	14	26	20	10	13	16	14	10	11	14	20	18	13	22	22	57.1	14.3	25.0	10.0
North Carolina	100	95	93	68	76	64	35	64	71	84	89	78	87	98	98	-2.0	-36.0	39.1	10.1
Oklahoma	34	29	20	13	22	18	19	13	13	18	28	27	28	29	24	-29.4	-47.1	55.6	-14.3
South Carolina	75	46	42	46	48	37	48	54	46	42	37	62	48	62	60	-20.0	-50.7	0.0	62.2
Tennessee	78	71	51	47	68	65	81	56	50	44	56	65	73	52	49	-37.2	-16.7	-13.8	-12.5
Texas	217	199	238	215	193	171	184	175	115	162	172	188	192	195	165	-24.0	-21.2	0.6	-4.1
Virginia	54	73	71	61	68	68	66	73	73	62	79	89	86	92	98	81.5	25.9	16.2	24.1
West Virginia	24	19	14	15	14	14	15	14	14	18	14	19	20	19	25	4.2	-41.7	0.0	78.6
West	595	588	536	499	448	427	452	529	517	475	583	552	580	576	667	12.1	-28.2	36.5	14.4
Alaska	0	0	0	1	1	0	0	0	2	1	0	5	3	1	1	([1])	([1])	([1])	([1])
Arizona	28	27	25	20	13	28	11	22	17	20	23	22	31	42	46	64.3	0.0	-17.9	100.0
California	328	338	275	303	241	206	249	311	297	274	346	285	294	282	324	-1.2	-37.2	68.0	-6.4
Colorado	52	46	54	45	48	42	38	42	40	34	29	55	53	56	72	38.5	-19.2	-31.0	148.3
Hawaii	3	3	3	2	5	4	1	3	0	4	0	2	1	1	2	-33.3	33.3	-100.0	([1])
Idaho	13	10	8	4	3	6	3	4	8	5	11	8	8	7	11	-15.4	-53.8	83.3	0.0

Table 3-3.—Continued

Region and state	1975-76	1976-77	1977-78	1978-79	1979-80	1980-81	1981-82	1982-83	1983-84	1984-85	1985-86	1986-87	1987-88	1988-89	1989-90	Percent change: 1975-76 to 1989-90	Percent change: 1975-76 to 1980-81	Percent change: 1980-81 to 1985-86	Percent change: 1985-86 to 1989-90
West Continued																			
Montana	8	8	13	4	8	10	10	15	16	13	13	20	14	13	18	125.0	25.0	30.0	38.5
Nevada	1	9	8	1	2	4	5	9	10	2	3	2	2	4	4	300.0	300.0	-25.0	33.3
New Mexico	24	29	28	21	21	21	22	19	21	29	29	25	18	19	27	12.5	-12.5	38.1	-6.9
Oregon	61	45	63	45	47	38	46	48	32	32	48	44	50	42	47	-23.0	-37.7	26.3	-2.1
Utah	32	35	20	24	19	21	21	11	20	23	31	30	44	43	53	65.6	-34.4	47.6	71.0
Washington	41	34	35	25	36	41	38	39	47	36	36	46	37	49	48	17.1	0.0	-12.2	33.3
Wyoming	4	4	4	4	4	6	8	6	7	2	14	8	25	17	14	250.0	50.0	133.3	0.0
U.S. Service Schools	3	0	4	3	1	2	2	3	0	2	2	0	0	1	0	-100.0	-33.3	0.0	-100.0
Outlying Areas	6	3	10	10	8	2	4	2	8	6	12	8	11	6	7	16.7	-66.7	500.0	-41.7
American Samoa	—	—	—	—	—	—	—	—	—	—	—	—	—	—	—	([1])	([1])	([1])	([1])
Guam	0	0	0	0	0	0	0	0	0	0	0	—	—	—	—	([1])	([1])	([1])	([1])
Northern Marianas	—	—	—	—	—	—	—	—	—	—	—	—	—	—	—	([1])	([1])	([1])	([1])
Puerto Rico	6	3	10	10	8	2	4	2	8	6	12	8	11	6	7	16.7	-66.7	500.0	-41.7
Trust Territories	—	—	—	—	—	—	—	—	—	—	—	—	—	—	—	([1])	([1])	([1])	([1])
Virgin Islands	0	—	0	0	0	0	0	0	0	0	0	—	—	—	—	([1])	([1])	([1])	([1])

[1]Insufficient data for calculating a percent change.

—Data not available.

SOURCE: U.S. Department of Education, National Center for Education Statistics, HEGIS, "Degrees and Other Formal Awards Conferred" surveys; and IPEDS,"Completions" surveys.

Table 3-4.—Doctor's degrees conferred in mathematics, by region and state: 1975-76 to 1989-90

Region and state	1975-76	1976-77	1977-78	1978-79	1979-80	1980-81	1981-82	1982-83	1983-84	1984-85	1985-86	1986-87	1987-88	1988-89	1989-90	Percent change 1975-76 to 1989-90	Percent change 1975-76 to 1980-81	Percent change 1980-81 to 1985-86	Percent change 1985-86 to 1989-90
United States . . .	856	823	805	730	724	728	681	698	695	699	742	725	750	866	915	6.9	-15.0	1.9	23.3
Northeast	281	276	269	247	230	233	209	216	211	233	244	220	225	266	279	-0.7	-17.1	4.7	14.3
Connecticut . .	25	14	8	14	15	13	14	11	9	11	10	3	13	23	23	-8.0	-48.0	-23.1	130.0
Maine	0	0	0	0	0	0	0	0	0	0	0	—	—	—	—	([1])	([1])	([1])	([1])
Massachusetts	55	72	67	56	53	56	36	51	43	55	45	59	43	67	45	-18.2	1.8	-19.6	0.0
New Hampshire	6	2	7	2	3	2	3	2	5	2	6	3	1	6	1	-83.3	-66.7	200.0	-83.3
New Jersey . . .	38	40	28	36	24	35	27	28	27	26	33	19	25	24	35	-7.9	-7.9	-5.7	6.1
New York	107	108	98	90	90	85	91	84	88	82	95	94	80	76	120	12.1	-20.6	11.8	26.3
Pennsylvania .	39	26	42	30	27	26	22	25	32	39	33	31	38	46	43	10.3	-33.3	26.9	30.3
Rhode Island .	10	13	18	18	18	16	16	15	7	18	22	11	25	24	12	20.0	60.0	37.5	-45.5
Vermont	1	1	1	1	0	0	0	0	0	0	0	—	—	—	0	-100.0	-100.0	([1])	([1])
Midwest	229	228	209	178	180	174	196	176	170	164	183	194	196	221	242	5.7	-24.0	5.2	32.2
Illinois	49	53	49	45	37	38	43	30	27	35	30	42	46	51	60	22.4	-22.4	-21.1	100.0
Indiana	24	31	23	22	14	25	22	28	22	18	25	27	16	23	26	8.3	4.2	0.0	4.0
Iowa	13	20	17	13	17	19	17	21	19	24	17	21	28	19	25	92.3	46.2	-10.5	47.1
Kansas	6	5	14	5	5	6	4	5	7	2	6	11	4	5	5	-16.7	0.0	0.0	-16.7
Michigan	37	40	21	18	27	27	19	25	31	22	23	21	27	37	30	-18.9	-27.0	-14.8	30.4
Minnesota	15	8	8	4	11	11	18	9	9	8	16	7	14	19	21	40.0	-26.7	45.5	31.3
Missouri	19	15	13	21	15	15	15	16	7	5	13	11	9	10	8	-57.9	-21.1	-13.3	-38.5
Nebraska	1	5	3	2	5	1	5	0	0	1	2	3	3	3	3	200.0	0.0	100.0	50.0
North Dakota .	0	0	0	0	0	0	0	0	0	0	0	—	—	—	2	([1])	([1])	([1])	([1])
Ohio	33	24	40	23	18	17	29	21	21	20	23	24	18	26	35	6.1	-48.5	35.3	52.2
South Dakota .	0	0	0	0	0	0	0	0	0	0	0	—	—	—	0	([1])	([1])	([1])	([1])
Wisconsin	32	27	21	25	31	15	24	21	27	29	28	27	31	28	27	-15.6	-53.1	86.7	-3.6
South	159	167	179	150	154	150	133	157	151	150	162	139	155	183	214	34.6	-5.7	8.0	32.1
Alabama	7	5	6	8	6	4	5	5	2	2	2	7	7	3	9	28.6	-42.9	-50.0	350.0
Arkansas	1	2	0	0	1	4	0	1	1	1	1	0	0	3	1	0.0	300.0	-75.0	0.0
Delaware	4	1	5	1	3	0	1	3	0	1	5	5	2	7	7	75.0	-100.0	([1])	40.0
District of Columbia . . .	4	13	6	5	9	8	10	12	10	9	13	2	9	8	10	150.0	100.0	62.5	-23.1
Florida	18	17	18	25	20	20	19	20	11	21	11	13	12	12	13	-27.8	11.1	-45.0	18.2
Georgia	12	16	10	10	8	11	7	6	7	5	4	14	5	8	10	-16.7	-8.3	-63.6	150.0
Kentucky	7	9	5	8	9	5	5	4	6	2	4	1	3	6	5	-28.6	-28.6	-20.0	25.0
Louisiana	11	8	11	7	5	3	7	5	7	9	5	7	11	13	13	18.2	-72.7	66.7	160.0
Maryland	15	13	14	12	17	11	14	16	18	25	20	12	21	20	23	53.3	-26.7	81.8	15.0
Mississippi . . .	4	0	2	0	1	0	0	0	1	0	1	0	3	0	2	-50.0	-100.0	([1])	100.0
North Carolina	21	14	19	18	14	11	12	18	23	15	23	11	23	15	26	23.8	-47.6	109.1	13.0
Oklahoma	5	10	12	8	6	6	3	8	8	9	5	1	0	6	4	-20.0	20.0	-16.7	-20.0
South Carolina	5	5	6	7	2	3	6	5	5	6	12	5	8	5	8	60.0	-40.0	300.0	-33.3
Tennessee . . .	14	6	6	5	11	16	2	7	7	4	7	11	5	7	9	-35.7	14.3	-56.3	28.6
Texas	23	36	43	27	30	37	34	28	31	24	32	35	35	51	50	117.4	60.9	-13.5	56.3
Virginia	8	12	16	9	12	11	8	19	14	17	17	15	11	19	24	200.0	37.5	54.5	41.2
West Virginia .	0	0	0	0	0	0	0	0	0	0	0	0	0	0	0	([1])	([1])	([1])	([1])
West	187	152	148	155	160	171	143	149	163	152	153	172	174	196	180	-3.7	-8.6	-10.5	17.6
Alaska	0	0	0	0	0	0	0	0	0	0	0	—	1	1	—	([1])	([1])	([1])	([1])
Arizona	3	5	3	5	4	3	1	8	10	4	8	8	4	9	12	300.0	0.0	166.7	50.0
California	119	96	96	105	101	109	93	89	104	105	95	99	100	110	101	-15.1	-8.4	-12.8	6.3
Colorado	21	13	14	12	15	12	19	14	15	8	8	13	17	16	15	-28.6	-42.9	-33.3	87.5
Hawaii	1	0	0	1	0	2	0	0	1	3	1	1	—	4	1	0.0	100.0	-50.0	0.0
Idaho	3	1	2	3	2	5	1	4	1	3	1	1	2	6	3	0.0	66.7	-80.0	200.0

Table 3-4.—Continued

Region and state	1975-76	1976-77	1977-78	1978-79	1979-80	1980-81	1981-82	1982-83	1983-84	1984-85	1985-86	1986-87	1987-88	1988-89	1989-90	Percent change: 1975-76 to 1989-90	Percent change: 1975-76 to 1980-81	Percent change: 1980-81 to 1985-86	Percent change: 1985-86 to 1989-90
West Continued																			
Montana	4	4	4	2	4	4	4	2	4	0	0	5	2	4	3	-25.0	0.0	-100.0	([1])
Nevada	0	0	0	0	0	0	0	0	0	0	0	—	—	—	—	([1])	([1])	([1])	([1])
New Mexico	7	10	6	1	3	8	4	2	5	6	3	12	8	8	9	28.6	14.3	-62.5	200.0
Oregon	12	9	13	14	11	12	7	13	7	9	13	8	10	9	14	16.7	0.0	8.3	7.7
Utah	4	8	3	6	5	3	4	3	2	2	6	10	10	7	4	0.0	-25.0	100.0	-33.3
Washington	11	5	5	6	14	9	10	14	13	11	16	15	19	20	16	45.5	-18.2	77.8	0.0
Wyoming	2	1	2	0	1	4	0	0	1	1	2	—	1	2	2	0.0	100.0	-50.0	0.0
U.S. Service Schools	0	0	0	0	0	0	0	0	0	0	0	0	0	0	0	([1])	([1])	([1])	([1])
Outlying Areas	0	0	0	0	0	0	0	0	0	0	0	0	0	0	0	([1])	([1])	([1])	([1])
American Samoa	—	—	—	—	—	—	—	—	—	—	—	—	—	—	—	([1])	([1])	([1])	([1])
Guam	0	0	0	0	0	0	0	0	0	0	0	—	—	—	—	([1])	([1])	([1])	([1])
Northern Marianas	—	—	—	—	—	—	—	—	—	—	—	—	—	—	—	([1])	([1])	([1])	([1])
Puerto Rico	0	0	0	0	0	0	0	0	0	0	0	—	—	0	0	([1])	([1])	([1])	([1])
Trust Territories	—	—	—	—	—	—	—	—	—	—	—	—	—	—	—	([1])	([1])	([1])	([1])
Virgin Islands	0	—	0	0	0	0	0	0	0	0	0	—	—	—	—	([1])	([1])	([1])	([1])

[1]Insufficient data for calculating a percent change.

—Data not available.

SOURCE: U.S. Department of Education, National Center for Education Statistics, HEGIS, "Degrees and Other Formal Awards Conferred" surveys; and IPEDS,"Completions" surveys.

THIS PAGE INTENTIONALLY LEFT BLANK

4. Supply of Graduates in Science: Associate, Bachelor's, Master's and Doctor's Degrees, by Region and State

Study of regional and state patterns in degrees conferred in science[1] is a necessary part of understanding the current supply, and possible future supply, of individuals with technical and scientific backgrounds. Chapter 2 presented changes as national averages; closer study reveals that although most trends are supported by most regions and states, some states and regions go against the trends, or have unusually large increases or decreases in numbers of degrees conferred that are worth noting.

Most degrees in science are conferred at the bachelor's level. Bachelor's degrees in science increased in every region of the country during the late 1970s and early 1980s. During the late 1980s, bachelor's degrees in science dropped in every region, while master's and doctor's in science rose in every region during that same period.

[1] Science is defined as agricultural sciences, computer sciences, engineering, health sciences, life sciences, and physical sciences.

Text table 2.—Science degrees by region and degree level: 1975-76 to 1989-90

Degree level and region	1975-76	1982-83	1985-86	1989-90	Percent change: 1982-83 to 1989-90	Percent change: 1982-83 to 1985-86	Percent change: 1985-86 to 1989-90
Associate degrees							
United States	—	147,777	149,448	133,864	-9.4	1.1	-10.4
Northeast	—	33,553	31,682	28,840	-14.0	-5.6	-9.0
Midwest	—	41,413	41,816	36,367	-12.2	1.0	-13.0
South	—	42,695	43,257	40,132	-6.0	1.3	-7.2
West	—	27,176	28,124	24,241	-10.8	3.5	-13.8
U.S.S.S	—	2,940	4,569	4,284	45.7	55.4	-6.2

Degree level and region	1975-76	1980-81	1985-86	1989-90	Percent change: 1975-76 to 1989-90	Percent change: 1975-76 to 1980-81	Percent change: 1980-81 to 1985-86	Percent change: 1985-86 to 1989-90
Bachelor's degrees								
United States	200,938	242,523	279,455	234,731	16.8	20.7	15.2	-16.0
Northeast	50,075	60,020	68,281	54,749	9.3	19.9	13.8	-19.8
Midwest	56,859	69,292	79,850	69,198	21.7	21.9	15.2	-13.3
South	56,373	69,974	80,758	67,224	19.2	24.1	15.4	-16.8
West	36,524	41,897	48,983	42,423	16.2	14.7	16.9	-13.4
U.S.S.S	1,107	1,340	1,583	1,137	2.7	21.0	18.1	-28.2
Master's degrees								
United States	46,218	52,196	63,071	68,526	48.3	12.9	20.8	8.6
Northeast	12,266	13,459	16,991	18,640	52.0	9.7	26.2	9.7
Midwest	11,831	13,132	15,322	16,157	36.6	11.0	16.7	5.4
South	12,083	14,119	17,563	18,769	55.3	16.9	24.4	6.9
West	9,587	11,002	12,591	14,650	52.8	14.8	14.4	16.4
U.S.S.S	451	484	604	310	-31.3	7.3	24.8	-48.7
Doctor's degrees								
United States	11,393	11,566	13,062	16,415	44.1	1.5	12.9	25.7
Northeast	3,005	2,974	3,285	4,092	36.2	-1.0	10.5	24.6
Midwest	3,179	3,212	3,443	4,223	32.8	1.0	7.2	22.7
South	2,643	2,634	3,274	4,311	63.1	-0.3	24.3	31.7
West	2,538	2,735	3,041	3,789	49.3	7.8	11.2	24.6
U.S.S.S	28	11	19	0	-100.0	-60.7	72.7	-100.0

Note: U.S.S.S.=U. S. Service Schools.

—Data not available.

SOURCE: U.S. Department of Education, National Center for Education Statistics, HEGIS, "Degrees and Other Formal Awards Conferred" surveys; and IPEDS,"Completions" surveys.

Associate Degrees

Regions

Associate degrees conferred in science declined in all regions between 1982–83[2] and 1989–90. Text table 2 shows that the largest decrease was in the Northeast (14 percent); the smallest was in the South (6 percent). The South and Midwest conferred the most, and the West consistently awarded the smallest numbers of associate degrees in science throughout the period.

States

Table 4–1 presents data on associates in science on a state-by-state basis. Despite a national decline in the awarding of associate degrees in science, 20 states showed no change or reported a gain in the numbers of these degrees between 1982–83 and 1989–90, especially Oklahoma, with an increase of 82 percent. The Service Schools also had a notable increase, 46 percent. Of the 11 states in which over 4,500 associate degrees were awarded in 1982–83, only one, Texas, had increased that number in 1989–90, all the others had decreases of 9 percent or more.

Bachelor's Degrees

Regions

Bachelor's degrees in science rose in every region in the Nation between 1975–76 and 1989–90 (text table 2). The largest increase was in the Midwest (22 percent), followed by the South (19 percent), the West (16 percent), and the Northeast (9 percent). Between 1975–76 and 1980–81, increases in numbers of bachelor's degrees awarded, by region, ranged from 15 percent in the West to 24 percent in the South. The increases were somewhat smaller from 1980–81 to 1985–86, in every region except the West.

The picture changed in the late 1980s. Between 1985–86 and 1989–90, bachelor's degrees in science declined in every region, 13 percent in the Midwest and West, 17 percent in the South, and 20 percent in the Northeast. The Midwest and South conferred the largest numbers of bachelor's in science in 1989–90, and the West the smallest number.

2 The academic year 1982–83 is the first year for which the field of study data are consistent for associate degrees.

Chart 3 — Bachelor's degrees in science, by region: 1975–76 to 1989–90

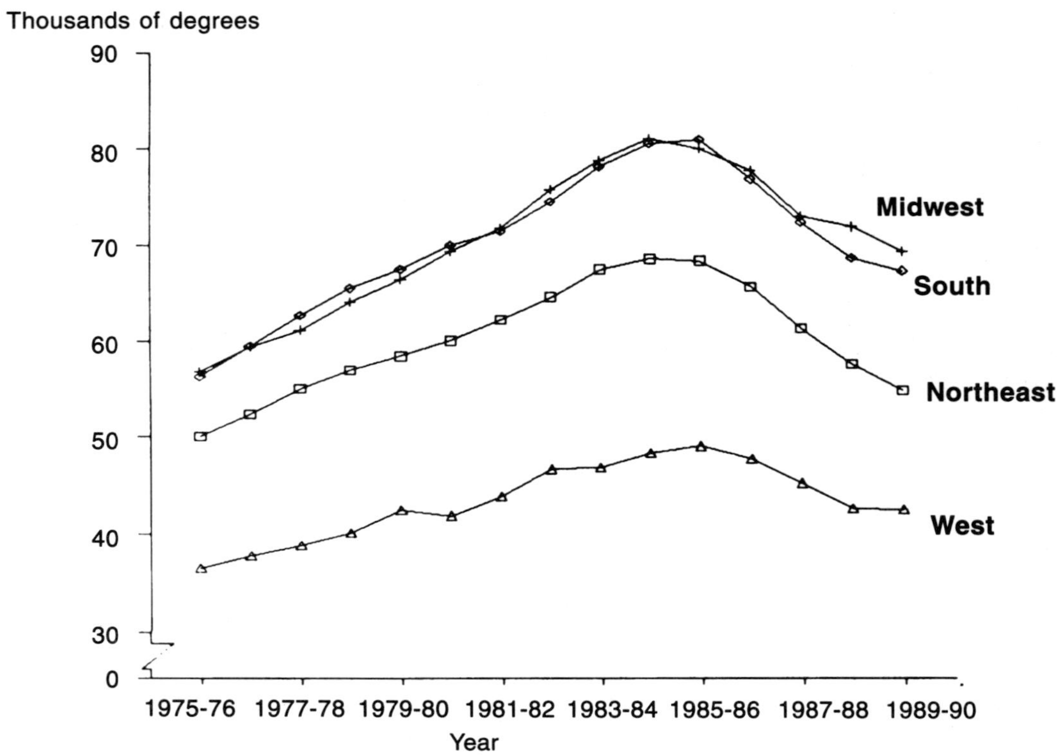

Source: U.S. Department of Education, National Center for Education Statistics, HEGIS, "Degrees and Other Formal Awards Conferred" surveys; and IPEDS, "Completions" surveys.

States

Only eight states in the country had declines in the numbers of science degrees at the bachelor's level between 1975–76 and 1989–90. Increases in bachelor's degrees in science ranged from less than 1 percent in Tennessee to 64 percent in Florida (table 4–2). Between 1975–76 and 1980–81, every state in the country had an increase in the number of bachelor's degrees conferred, and between 1980–81 and 1985–86 only one state, Washington, had a slight decline (1 percent).

The late 1980s showed a dramatic change in this pattern of widespread increases. Bachelor's degrees awarded in science between 1985–86 and 1989–90 fell in every state in the country, the District of Columbia, and the Service Schools. The decreases ranged from 2 percent in Wisconsin to 29 percent in Rhode Island and West Virginia. In the Northeast the decline was at least 15 percent in every state. In 1989–90, California produced the largest number of bachelor's degrees in science (22,400, almost 10 percent of the total), having overtaken New York in the early 1980s.

Science Degrees as a Proportion of Total Degrees

Table 4–3 presents bachelor's degrees in science as a percentage of all bachelor's degrees, by region and state. In 1975–76, bachelor's degrees in science represented 22 percent of all bachelor's degrees on the national level. The percentage of bachelor's degrees in science rose each year until 1984–85 when it peaked at 29 percent, and then dropped each year reaching 22 percent in 1989–90.

On the regional level, the Midwest led the country in the proportion of bachelor's degrees awarded to graduates in science throughout the period, going from 23 percent in 1975–76 to 30 percent in 1984–85, and then dropping to 24 percent in 1989–90. During 1989–90, North Dakota was the state that awarded the largest proportion of bachelor's degrees in science, over 31 percent, followed by South Dakota, Montana, Maine, Wisconsin, Indiana, Michigan, Wyoming, Illinois, New Mexico, and Colorado, all awarding over 25 percent of their bachelor's degrees to science graduates.

Most of the states in the South and Midwest increased their share of bachelor's degrees in science between 1975–76 and 1989–90, while this share of degrees decreased in about half the states in the Northeast and most states in the West over the same period. There were some big proportional changes within individual states in both directions, such as gains of over 4 percentage points in Maine, Illinois, and Florida, and decreases of 7 percentage points or more in Rhode Island, Delaware, Alaska, and Wyoming.

In 1989–90, about a quarter of the states conferred less than 21 percent of their bachelor's degrees in science, with Rhode Island (15 percent) having the lowest proportion of bachelor's degrees going to science graduates. There was also considerable variation within the regions. Except for the South, there was a range of more than 10 percentage points among states, for example in the Midwest, Minnesota conferred 18 percent of its bachelor's degrees in science, while in North Dakota (as mentioned previously) the percentage was over 31.

Master's Degrees

Regions

Between 1975–76 and 1989–90, master's degrees in science rose almost every year in all four regions. Text table 2 shows that over the entire period, the numbers of master's degrees in science rose 37 percent in the Midwest and over 50 percent in the Northeast, South, and West. Increases were the largest in the early 1980s; between 1980–81 and 1985–86 master's degrees in science rose 14 percent in the West, 17 percent in the Midwest, 24 percent in the South, and 26 percent in the Northeast. Master's degrees in science had smaller increases between 1985–86 and 1989–90, with the exception of the West. In 1989–90, the Northeast and South conferred the highest numbers of master's degrees in science, while the West awarded the lowest amount.

States

Master's degrees in science rose in 47 states and the District of Columbia between 1975–76 and 1989–90, with increases of over 100 percent in New Hampshire, Alabama, Florida, Maryland, Virginia, and Alaska (table 4–4). However, in the late 1980s, master's degrees awarded in science dropped in 13 states and the District of Columbia. In 1989–90, California and New York awarded the most master's degrees in science, and these two states combined accounted for about 23 percent of all of the master's degrees conferred in science.

Doctor's Degrees

Regions

Between 1975–76 and 1989–90, increases in the numbers of doctor's degrees conferred in science ranged from 33 percent in the Midwest to 63 percent in the South (text table 2). In contrast to the declining numbers of bachelor's degrees and relatively

Chart 4 — Bachelor's degrees in science as a percent of all bachelor's degrees, by region: 1975–76 to 1989–90

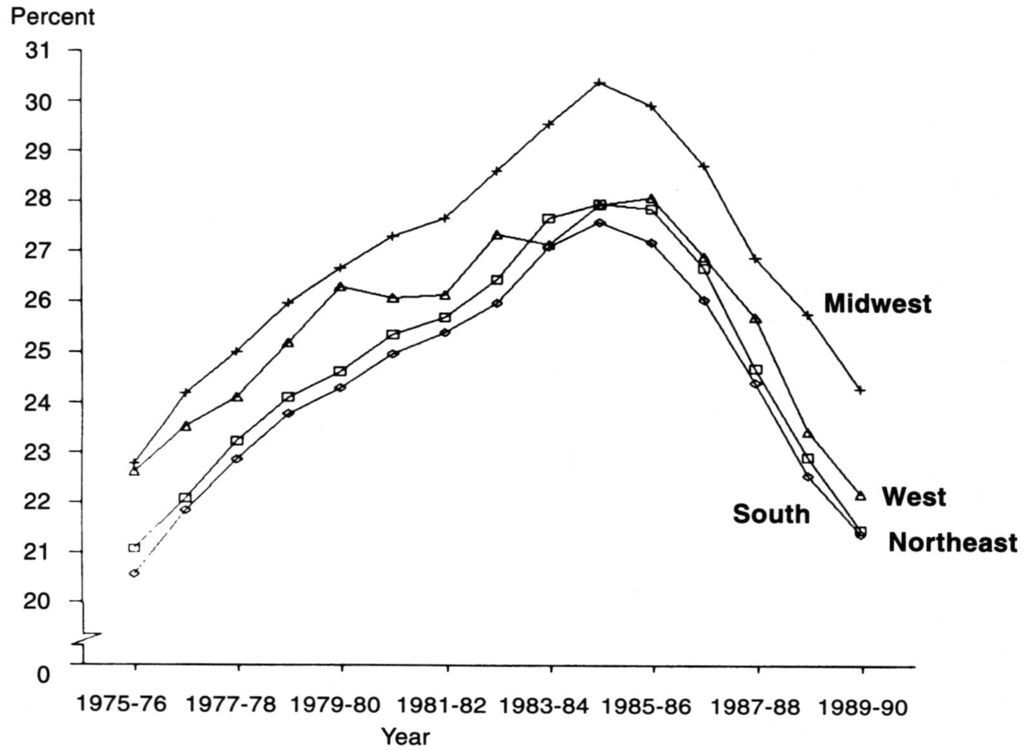

Source: U.S. Department of Education, National Center for Education Statistics, HEGIS, "Degrees and Other Formal Awards Conferred" surveys; and IPEDS, "Completions" surveys.

small increases in master's degrees between 1985–86 and 1989–90, doctor's degrees increased substantially. These increases ranged from 23 percent in the Midwest to 25 percent in the Northeast and West to 32 percent in the South.

The South moved from being third among the regions in 1975–76 in doctor's degrees awarded in science, to first in 1989–90, because of growth in the numbers of these degrees awarded during the 1980s.

States

Between 1975–76 and 1989–90 the numbers of doctor's degrees in science rose in every state, except Oklahoma, West Virginia, and Alaska. Table 4–5 shows that increases were over 100 percent in Alabama, Virginia, Nevada, and New Mexico. Doctor's degrees awarded in science between 1985–86 and 1989–90 rose in 44 states with the increases ranging from 3 percent in Montana and Oregon to 165 percent in Idaho. Only three states conferred at least 1,000 doctor's degrees in science in 1989–90; California (2,200), New York (1,500), and Texas (1,000). These three states accounted for 29 percent of all the doctor's degrees awarded in science in 1989–90.

Summary

A look at science degrees by region and state shows that trends differ depending on the level of the degree. Master's and doctor's degrees increased in all areas of the country in the 1980s. Associate and bachelor's degrees in science, the building blocks for advanced study in this area, were being awarded in decreasing numbers throughout the country in the late 1980s. Bachelor's degrees in science have also been declining as a percentage of all bachelor's degrees.

One area for future study is the accessibility of science programs in the colleges and universities. Another factor that may need to be examined is the effect the decline in bachelor's degrees conferred in science during the late 1980s may have on graduate study in science in the 1990s and beyond.

Table 4-1.—Associate degrees conferred in science, by region and state: 1982-83 to 1989-90

Region and state	1975-76	1976-77	1977-78	1978-79	1979-80	1980-81	1981-82	1982-83	1983-84	1984-85	1985-86	1986-87	1987-88	1988-89	1989-90	Percent change 1982-83 to 1989-90	Percent change 1982-83 to 1985-86	Percent change 1985-86 to 1989-90
United States	—	—	—	—	—	—	—	147,777	153,390	154,830	149,448	142,571	138,339	131,502	133,864	-9.4	1.1	-10.4
Northeast	—	—	—	—	—	—	—	33,553	34,176	33,243	31,682	29,772	29,544	27,337	28,840	-14.0	-5.6	-9.0
Connecticut	—	—	—	—	—	—	—	1,582	1,618	1,652	1,422	1,335	1,315	1,281	1,313	-17.0	-10.1	-7.7
Maine	—	—	—	—	—	—	—	793	774	924	691	755	757	629	626	-21.1	-12.9	-9.4
Massachusetts	—	—	—	—	—	—	—	5,222	5,291	4,922	4,581	4,072	3,910	3,724	3,710	-29.0	-12.3	-19.0
New Hampshire	—	—	—	—	—	—	—	915	1,129	1,260	1,156	1,011	931	899	897	-2.0	26.3	-22.4
New Jersey	—	—	—	—	—	—	—	3,008	3,111	3,024	2,991	2,643	2,472	2,297	2,412	-19.8	-0.6	-19.4
New York	—	—	—	—	—	—	—	12,858	12,569	12,217	11,664	11,209	11,724	11,336	11,734	-8.7	-9.3	0.6
Pennsylvania	—	—	—	—	—	—	—	7,698	8,100	7,701	7,589	7,033	7,124	5,969	6,633	-13.8	-1.4	-12.6
Rhode Island	—	—	—	—	—	—	—	962	1,084	1,014	1,114	1,313	944	881	1,088	13.1	15.8	-2.3
Vermont	—	—	—	—	—	—	—	515	500	529	474	401	367	321	427	-17.1	-8.0	-9.9
Midwest	—	—	—	—	—	—	—	41,413	44,374	44,712	41,816	41,080	39,519	37,534	36,367	-12.2	1.0	-13.0
Illinois	—	—	—	—	—	—	—	6,638	6,833	6,870	6,808	6,641	6,809	5,862	4,818	-27.4	2.6	-29.2
Indiana	—	—	—	—	—	—	—	4,540	4,954	4,956	4,606	4,396	4,410	4,128	4,145	-8.7	1.5	-10.0
Iowa	—	—	—	—	—	—	—	2,462	2,690	2,926	2,785	2,846	2,699	3,049	3,131	27.2	13.1	12.4
Kansas	—	—	—	—	—	—	—	1,409	1,615	1,590	1,389	1,397	1,280	1,411	1,608	14.1	-1.4	15.8
Michigan	—	—	—	—	—	—	—	7,930	8,440	8,475	7,627	7,613	6,792	6,023	6,066	-23.5	-3.8	-20.5
Minnesota	—	—	—	—	—	—	—	2,165	2,195	2,248	1,839	2,225	2,173	1,885	1,934	-10.7	-15.1	5.2
Missouri	—	—	—	—	—	—	—	1,785	2,148	2,349	2,125	2,053	2,070	1,977	1,707	-4.4	19.0	-19.7
Nebraska	—	—	—	—	—	—	—	1,614	1,618	1,546	1,348	1,307	1,200	1,115	1,169	-27.6	-16.5	-13.3
North Dakota	—	—	—	—	—	—	—	826	766	712	820	848	777	655	698	-15.5	-0.7	-14.9
Ohio	—	—	—	—	—	—	—	8,511	8,939	9,182	8,715	8,396	7,994	8,266	7,751	-8.9	2.4	-11.1
South Dakota	—	—	—	—	—	—	—	613	652	476	467	352	369	342	342	-44.2	-23.8	-26.8
Wisconsin	—	—	—	—	—	—	—	2,920	3,524	3,382	3,287	3,006	2,946	2,821	2,998	2.7	12.6	-8.8
South	—	—	—	—	—	—	—	42,695	42,837	44,385	43,257	40,140	39,212	39,580	40,132	-6.0	1.3	-7.2
Alabama	—	—	—	—	—	—	—	1,688	1,951	2,613	2,575	2,620	2,347	2,165	2,171	28.6	52.5	-15.7
Arkansas	—	—	—	—	—	—	—	917	996	978	920	1,127	1,178	1,152	1,138	24.1	0.3	23.7
Delaware	—	—	—	—	—	—	—	581	608	587	546	541	523	499	584	0.5	-6.0	7.0
District of Columbia	—	—	—	—	—	—	—	277	278	339	276	227	237	237	226	-18.4	-0.4	-18.1
Florida	—	—	—	—	—	—	—	7,494	6,466	6,712	7,133	6,105	5,638	5,674	5,410	-27.8	-4.8	-24.2
Georgia	—	—	—	—	—	—	—	1,917	2,374	2,144	2,179	2,178	2,015	2,268	2,171	13.2	13.7	-0.4
Kentucky	—	—	—	—	—	—	—	2,599	2,837	3,142	2,694	2,217	2,058	2,056	2,078	-20.0	3.7	-22.9
Louisiana	—	—	—	—	—	—	—	966	1,056	1,124	1,179	1,085	971	976	893	-7.6	22.0	-24.3
Maryland	—	—	—	—	—	—	—	1,762	1,862	1,844	1,750	1,666	1,561	1,384	1,534	-12.9	-0.7	-12.3
Mississippi	—	—	—	—	—	—	—	2,050	1,964	2,006	1,815	1,663	1,602	1,653	1,611	-21.4	-11.5	-11.2
North Carolina	—	—	—	—	—	—	—	3,466	3,942	3,743	3,675	3,514	3,289	3,185	3,501	1.0	6.0	-4.7
Oklahoma	—	—	—	—	—	—	—	1,327	1,672	1,733	2,408	2,100	1,943	2,522	2,411	81.7	81.5	0.1
South Carolina	—	—	—	—	—	—	—	1,686	1,909	1,844	1,809	1,628	1,641	1,660	1,790	6.2	7.3	-1.1
Tennessee	—	—	—	—	—	—	—	2,996	2,878	2,910	2,673	2,469	2,543	2,317	2,375	-20.7	-10.8	-11.1
Texas	—	—	—	—	—	—	—	7,108	8,090	8,540	7,825	7,377	7,934	8,016	8,210	15.5	10.1	4.9
Virginia	—	—	—	—	—	—	—	4,573	2,601	2,561	2,332	2,360	2,608	2,499	2,700	-41.0	-49.0	15.8
West Virginia	—	—	—	—	—	—	—	1,288	1,353	1,565	1,468	1,263	1,124	1,317	1,329	3.2	14.0	-9.5
West	—	—	—	—	—	—	—	27,176	29,063	29,163	28,124	26,962	26,179	24,071	24,241	-10.8	3.5	-13.8
Alaska	—	—	—	—	—	—	—	259	218	214	189	211	178	163	151	-41.7	-27.0	-20.1
Arizona	—	—	—	—	—	—	—	2,104	2,018	2,190	2,550	1,578	2,150	2,312	1,932	-8.2	21.2	-24.2
California	—	—	—	—	—	—	—	13,727	15,004	15,347	13,253	13,389	12,374	10,984	10,708	-22.0	-3.5	-19.2
Colorado	—	—	—	—	—	—	—	2,124	2,132	2,184	2,192	2,738	2,638	2,287	2,572	21.1	3.2	17.3
Hawaii	—	—	—	—	—	—	—	646	649	624	643	610	561	519	505	-21.8	-0.5	-21.5
Idaho	—	—	—	—	—	—	—	763	807	835	762	691	775	799	899	17.8	-0.1	18.0

Table 4-1.—Continued

Region and state	1975-76	1976-77	1977-78	1978-79	1979-80	1980-81	1981-82	1982-83	1983-84	1984-85	1985-86	1986-87	1987-88	1988-89	1989-90	Percent change: 1982-83 to 1989-90	Percent change: 1982-83 to 1985-86	Percent change: 1985-86 to 1989-90
West Continued																		
Montana	—	—	—	—	—	—	—	238	276	272	238	226	254	208	247	3.8	0.0	3.8
Nevada	—	—	—	—	—	—	—	298	395	317	346	275	311	298	248	-16.8	16.1	-28.3
New Mexico	—	—	—	—	—	—	—	829	934	801	787	708	685	577	848	2.3	-5.1	7.8
Oregon	—	—	—	—	—	—	—	2,179	2,286	2,340	2,242	2,038	1,945	1,918	1,871	-14.1	2.9	-16.5
Utah	—	—	—	—	—	—	—	1,038	1,081	1,024	940	978	1,133	1,018	1,038	0.0	-9.4	10.4
Washington	—	—	—	—	—	—	—	2,600	2,741	2,536	3,491	3,089	2,548	2,463	2,735	5.2	34.3	-21.7
Wyoming	—	—	—	—	—	—	—	371	522	479	491	431	627	525	487	31.3	32.3	-0.8
U.S. Service Schools	—	—	—	—	—	—	—	2,940	2,940	3,327	4,569	4,617	3,885	2,980	4,284	45.7	55.4	-6.2
Outlying Areas	—	—	—	—	—	—	—	1,683	2,097	1,497	1,702	1,491	1,448	1,463	1,427	-15.2	1.1	-16.2
American Samoa	—	—	—	—	—	—	—	20	32	32	13	0	0	0	0	-100.0	-35.0	-100.0
Guam	—	—	—	—	—	—	—	22	45	27	35	30	37	39	32	45.5	59.1	-8.6
Northern Marianas	—	—	—	—	—	—	—	0	0	0	0	0	0	0	1	([1])	([1])	([1])
Puerto Rico	—	—	—	—	—	—	—	1,630	2,000	1,407	1,632	1,457	1,393	1,405	1,381	-15.3	0.1	-15.4
Trust Territories	—	—	—	—	—	—	—	0	12	12	8	0	10	10	10	([1])	([1])	25.0
Virgin Islands	—	—	—	—	—	—	—	11	8	19	14	4	8	9	3	-72.7	27.3	-78.6

[1] Insufficient data for calculating a percent change.

—Data not available.

SOURCE: U.S. Department of Education, National Center for Education Statistics, HEGIS, "Degrees and Other Formal Awards Conferred" surveys; and IPEDS,"Completions" surveys.

Table 4-2.—Bachelor's degrees conferred in science, by region and state: 1975-76 to 1989-90

Region and state	1975-76	1976-77	1977-78	1978-79	1979-80	1980-81	1981-82	1982-83	1983-84	1984-85	1985-86	1986-87	1987-88	1988-89	1989-90	Percent change: 1975-76 to 1989-90	Percent change: 1975-76 to 1980-81	Percent change: 1980-81 to 1985-86	Percent change: 1985-86 to 1989-90
United States	200,938	210,381	219,161	228,100	236,236	242,523	250,377	262,690	272,582	279,780	279,455	269,080	252,766	241,554	234,731	16.8	20.7	15.2	-16.0
Northeast	50,075	52,434	55,082	57,003	58,450	60,020	62,207	64,577	67,421	68,531	68,281	65,577	61,278	57,529	54,749	9.3	19.9	13.8	-19.8
Connecticut	2,675	2,905	3,071	3,021	3,012	3,091	2,878	3,031	3,040	3,244	3,336	3,060	2,889	2,567	2,544	-4.9	15.6	7.9	-23.7
Maine	1,054	1,033	1,111	1,134	1,441	1,429	1,339	1,594	1,574	1,552	1,666	1,747	1,653	1,492	1,382	31.1	35.6	16.6	-17.0
Massachusetts	8,411	8,856	9,438	9,320	9,444	9,717	10,311	10,692	11,896	11,801	11,831	11,608	10,780	10,054	9,416	11.9	15.5	21.8	-20.4
New Hampshire	1,119	1,184	1,254	1,245	1,343	1,348	1,452	1,657	1,447	1,616	1,582	1,469	1,304	1,311	1,201	7.3	20.5	17.4	-24.1
New Jersey	4,849	5,353	5,534	5,635	5,484	5,909	6,013	6,450	6,214	6,384	6,320	6,114	5,469	5,175	4,818	-0.6	21.9	7.0	-23.8
New York	18,135	18,417	19,436	20,141	20,643	20,729	21,595	21,680	22,998	23,091	22,914	21,963	20,507	19,785	18,441	1.7	14.3	10.5	-19.5
Pennsylvania	11,522	12,166	12,783	13,818	14,477	15,108	15,743	16,808	17,478	18,043	17,782	16,965	16,199	14,845	14,783	28.3	31.1	17.7	-16.9
Rhode Island	1,467	1,601	1,618	1,791	1,651	1,795	1,900	1,768	1,882	1,785	1,846	1,782	1,628	1,492	1,309	-10.8	22.4	2.8	-29.1
Vermont	843	919	837	898	955	894	976	897	892	1,015	1,004	869	849	808	855	1.4	6.0	12.3	-14.8
Midwest	56,859	59,483	61,182	64,104	66,494	69,292	71,598	75,673	78,618	80,816	79,850	77,570	72,831	71,702	69,198	21.7	21.9	15.2	-13.3
Illinois	9,483	10,463	10,559	11,511	12,184	12,812	12,984	13,105	14,095	14,230	14,335	14,077	13,558	13,086	12,588	32.7	35.1	11.9	-12.2
Indiana	6,092	6,256	6,378	6,528	7,033	7,070	7,357	8,210	8,229	8,556	8,565	8,283	7,631	7,420	7,293	19.7	16.1	21.1	-14.9
Iowa	2,860	2,837	2,974	3,316	3,374	3,688	3,598	3,766	3,877	4,101	4,302	3,976	3,833	3,585	3,326	16.3	29.0	16.6	-22.7
Kansas	2,787	2,997	3,069	3,296	3,347	3,293	3,641	3,889	4,016	3,861	3,704	3,387	3,105	3,086	3,082	10.6	18.2	12.5	-16.8
Michigan	9,405	9,628	10,028	10,417	10,767	11,526	11,756	12,301	12,623	12,675	12,416	12,074	11,606	11,426	11,192	19.0	22.6	7.7	-9.9
Minnesota	3,916	3,930	3,852	4,111	4,246	4,232	4,426	4,750	4,943	4,959	4,538	4,570	4,408	4,183	4,019	2.6	8.1	7.2	-11.4
Missouri	4,686	5,057	5,207	5,440	5,502	5,786	6,116	6,603	6,728	7,425	6,807	6,640	6,183	5,954	5,888	25.7	23.5	17.6	-13.5
Nebraska	1,767	1,721	2,009	1,783	1,836	1,818	1,924	1,986	2,064	2,159	2,163	2,132	2,029	1,842	1,759	-0.5	2.9	19.0	-18.7
North Dakota	1,059	1,119	1,125	1,212	1,246	1,273	1,280	1,408	1,525	1,590	1,652	1,540	1,440	1,420	1,321	24.7	20.2	29.8	-20.0
Ohio	8,542	8,944	9,011	9,422	9,531	10,193	10,464	11,179	11,653	12,473	12,627	12,138	11,190	11,330	10,515	23.1	19.3	23.9	-16.7
South Dakota	1,077	1,176	1,222	1,145	1,275	1,190	1,371	1,479	1,529	1,574	1,478	1,369	1,207	1,199	1,107	2.8	10.5	24.2	-25.1
Wisconsin	5,185	5,355	5,748	5,923	6,153	6,411	6,681	6,997	7,336	7,213	7,263	7,384	6,641	7,171	7,108	37.1	23.6	13.3	-2.1
South	56,373	59,557	62,723	65,534	67,529	69,974	71,335	74,434	78,016	80,387	80,758	76,705	72,210	68,546	67,224	19.2	24.1	15.4	-16.8
Alabama	3,100	3,260	3,559	3,811	3,997	4,250	4,403	4,530	4,720	4,880	4,706	4,651	4,507	4,310	4,177	34.7	37.1	10.7	-11.2
Arkansas	1,232	1,352	1,303	1,449	1,530	1,592	1,711	1,831	2,044	2,028	1,936	1,679	1,653	1,574	1,585	28.7	29.2	21.6	-18.1
Delaware	789	879	934	972	909	897	898	994	980	974	914	874	859	746	743	-5.8	13.7	1.9	-18.7
District of Columbia	1,254	1,290	1,408	1,457	1,407	1,514	1,514	1,624	1,668	1,680	1,723	1,692	1,520	1,558	1,473	17.5	20.7	13.8	-14.5
Florida	4,583	5,303	5,492	6,004	5,865	6,525	6,508	6,898	7,416	8,347	8,533	7,869	7,680	7,488	7,535	64.4	42.4	30.8	-11.7
Georgia	3,917	4,009	4,177	4,280	4,481	4,643	4,774	5,099	5,148	5,384	5,146	5,246	5,002	4,614	5,013	28.0	18.5	10.8	-2.6
Kentucky	2,347	2,458	2,632	2,669	2,896	2,837	2,963	3,058	3,269	3,358	3,325	3,140	3,060	2,849	2,817	20.0	20.9	17.2	-15.3
Louisiana	3,678	3,826	3,982	3,992	4,226	4,269	4,566	4,841	5,014	5,196	5,310	4,909	4,654	4,181	3,807	3.5	16.1	24.4	-28.3
Maryland	3,052	3,036	3,236	3,160	3,342	3,449	3,507	3,744	3,833	3,839	4,154	4,028	3,986	3,910	3,786	24.0	13.0	20.4	-8.9
Mississippi	1,849	1,970	1,881	2,087	2,182	2,304	2,206	2,414	2,488	2,529	2,668	2,662	2,233	2,075	1,983	7.2	24.6	15.8	-25.7
North Carolina	4,487	4,780	5,110	5,034	5,238	5,445	5,579	5,706	6,158	6,122	6,258	5,914	5,697	5,405	5,483	22.2	21.4	14.9	-12.4
Oklahoma	2,904	2,892	2,999	3,243	3,245	3,522	3,274	3,535	3,723	3,934	3,857	3,687	3,215	3,177	3,186	9.7	21.3	9.5	-17.4
South Carolina	2,113	2,328	2,490	2,401	2,553	2,451	2,373	2,512	2,640	2,850	3,005	2,879	2,599	2,878	2,492	17.9	16.0	22.6	-17.1
Tennessee	3,838	4,066	4,085	4,344	4,415	4,532	4,774	4,693	4,943	5,019	5,086	4,691	4,354	3,958	3,851	0.3	18.1	12.2	-24.3
Texas	11,537	12,201	12,769	13,671	14,176	14,458	14,653	15,026	15,545	15,614	15,628	14,748	13,574	12,664	12,437	7.8	25.3	8.1	-20.4
Virginia	4,157	4,257	4,747	5,111	5,144	5,217	5,543	5,720	6,128	6,367	6,267	6,002	5,916	5,490	5,254	26.4	25.5	20.1	-16.2
West Virginia	1,536	1,650	1,919	1,849	1,923	2,069	2,089	2,209	2,299	2,266	2,242	2,034	1,701	1,669	1,602	4.3	34.7	8.4	-28.5
West	36,524	37,778	38,873	40,107	42,420	41,897	43,846	46,655	46,857	48,284	48,983	47,646	45,173	42,526	42,423	16.2	14.7	16.9	-13.4
Alaska	140	117	114	102	154	150	169	243	201	255	277	272	256	245	216	54.3	7.1	84.7	-22.0
Arizona	2,263	2,481	2,453	2,548	2,831	2,859	3,181	3,489	3,527	3,927	3,850	3,679	3,398	3,084	2,902	28.2	26.3	34.7	-24.6
California	17,833	18,129	19,174	19,826	20,969	20,687	21,528	23,050	23,435	23,878	24,421	24,276	22,926	21,594	22,353	25.3	16.0	18.0	-8.5
Colorado	3,585	3,770	3,768	3,877	4,276	4,251	4,427	4,681	4,632	4,715	4,772	4,350	4,138	4,130	4,110	14.6	18.6	12.3	-13.9
Hawaii	540	589	686	641	686	658	699	620	699	654	734	752	725	630	630	16.7	21.9	11.6	-14.2
Idaho	694	785	778	793	934	812	852	946	946	890	890	819	743	687	728	4.9	17.0	9.6	-18.2
Montana	1,284	1,278	1,305	1,430	1,364	1,384	1,494	1,535	1,635	1,712	1,599	1,491	1,410	1,170	1,153	-10.2	7.8	15.5	-27.9
Nevada	351	325	366	335	394	373	363	438	445	451	464	456	416	406	411	17.1	6.3	24.4	-11.4
New Mexico	1,150	1,154	1,242	1,310	1,395	1,307	1,313	1,412	1,387	1,523	1,561	1,494	1,465	1,319	1,264	9.9	13.7	19.4	-19.0

Table 4-2.—Continued

Region and state	1975-76	1976-77	1977-78	1978-79	1979-80	1980-81	1981-82	1982-83	1983-84	1984-85	1985-86	1986-87	1987-88	1988-89	1989-90	Percent change 1975-76 to 1989-90	1975-76 to 1980-81	1980-81 to 1985-86	1985-86 to 1989-90
West Continued																			
Oregon	2,285	2,568	2,489	2,416	2,480	2,410	2,617	2,792	2,516	2,586	2,851	2,624	2,637	2,481	2,487	8.8	5.5	18.3	-12.8
Utah	1,952	2,226	1,965	2,283	2,305	2,353	2,353	2,560	2,648	2,756	2,920	2,809	2,610	2,454	2,242	14.9	20.5	24.1	-23.2
Washington	3,961	3,856	3,983	4,026	4,105	4,109	4,312	4,397	4,210	4,330	4,058	4,065	3,926	3,845	3,497	-11.7	3.7	-1.2	-13.8
Wyoming	486	500	550	520	527	544	538	492	576	607	586	559	523	481	430	-11.5	11.9	7.7	-26.6
U.S. Service Schools	1,107	1,129	1,301	1,352	1,343	1,340	1,391	1,351	1,670	1,762	1,583	1,582	1,274	1,251	1,137	2.7	21.0	18.1	-28.2
Outlying Areas	1,625	1,703	1,945	2,111	2,191	2,728	2,729	2,991	3,096	3,086	3,348	3,410	3,208	3,257	3,205	97.2	67.9	22.7	-4.3
American Samoa	0	0	0	0	0	0	0	0	0	0	0	0	0	0	0	([1])	([1])	([1])	([1])
Guam	10	6	12	8	5	9	14	4	11	7	9	9	12	15	10	0.0	-10.0	0.0	11.1
Northern Marianas	22	8	3	6	0	0	0	0	0	0	0	0	0	0	0	-100.0	-100.0	([1])	([1])
Puerto Rico	1,588	1,681	1,926	2,092	2,179	2,712	2,711	2,977	3,078	3,067	3,335	3,388	3,181	3,230	3,181	100.3	70.8	23.0	-4.6
Trust Territories	0	0	0	0	0	0	0	0	0	0	0	0	0	0	0	([1])	([1])	([1])	([1])
Virgin Islands	5	8	4	5	7	7	4	10	7	12	4	13	15	12	14	180.0	40.0	-42.9	250.0

[1] Insufficient data for calculating a percent change.

—Data not available.

SOURCE: U.S. Department of Education, National Center for Education Statistics, HEGIS, "Degrees and Other Formal Awards Conferred" surveys; and IPEDS,"Completions" surveys.

Table 4-3.—Bachelor's degrees conferred in science, as a percentage of all bachelor's degrees, by region and state: 1975-76 to 1989-90

Region and state	1975-76	1976-77	1977-78	1978-79	1979-80	1980-81	1981-82	1982-83	1983-84	1984-85	1985-86	1986-87	1987-88	1988-89	1989-90	Change in percentage points			
																1975-76 to 1989-90	1975-76 to 1980-81	1980-81 to 1985-86	1985-86 to 1989-90
United States . . .	21.71	22.88	23.79	24.76	25.42	25.93	26.27	27.10	27.98	28.56	28.29	27.14	25.41	23.71	22.36	0.7	4.2	2.4	-5.9
Northeast	21.07	22.09	23.22	24.10	24.61	25.34	25.68	26.45	27.66	27.94	27.84	26.67	24.68	22.90	21.46	0.4	4.3	2.5	-6.4
Connecticut . .	19.63	21.69	22.86	23.79	23.15	23.22	21.78	23.10	22.84	24.00	23.80	22.67	21.12	18.98	17.94	-1.7	3.6	0.6	-5.9
Maine	22.35	22.46	23.74	24.87	29.72	29.67	28.62	32.67	32.24	30.92	32.17	34.11	31.99	28.84	27.95	5.6	7.3	2.5	-4.2
Massachusetts	22.45	23.68	24.56	24.65	24.67	25.05	25.83	27.30	29.63	29.17	29.30	27.93	25.79	23.66	21.62	-0.8	2.6	4.2	-7.7
New Hampshire	21.30	22.34	20.30	22.98	23.35	22.37	23.76	24.73	22.91	25.60	24.12	21.70	19.17	18.87	17.81	-3.5	1.1	1.7	-6.3
New Jersey . . .	18.79	20.96	22.06	22.33	22.45	24.14	24.16	25.29	25.82	26.86	26.95	26.21	24.50	22.60	21.08	2.3	5.4	2.8	-5.9
New York	21.20	21.21	22.81	23.53	24.25	24.74	25.07	25.10	26.41	26.36	26.28	25.35	23.31	22.34	20.70	-0.5	3.5	1.5	-5.6
Pennsylvania .	21.11	22.62	23.89	25.49	26.32	27.75	28.37	29.50	31.02	31.73	31.38	29.68	27.76	25.21	24.44	3.3	6.6	3.6	-6.9
Rhode Island .	22.59	24.46	23.59	24.92	23.39	24.71	25.33	23.30	24.44	23.38	23.64	23.03	20.52	17.57	14.89	-7.7	2.1	-1.1	-8.7
Vermont	20.41	22.97	22.21	23.74	24.07	22.51	22.78	23.09	22.86	24.90	25.11	21.30	19.87	19.27	18.93	-1.5	2.1	2.6	-6.2
Midwest	22.78	24.17	25.00	25.98	26.66	27.29	27.67	28.60	29.54	30.38	29.89	28.71	26.86	25.76	24.26	1.5	4.5	2.6	-5.6
Illinois	21.17	23.15	24.21	26.39	27.48	28.69	28.62	28.23	29.61	30.57	30.23	29.71	28.27	26.78	25.30	4.1	7.5	1.5	-4.9
Indiana	25.30	26.29	27.02	27.57	29.03	28.47	28.65	31.02	31.65	32.42	32.17	31.11	28.90	27.61	26.40	1.1	3.2	3.7	-5.8
Iowa	21.31	21.74	22.53	24.79	24.97	25.54	24.88	25.98	26.10	26.86	27.15	24.17	22.89	21.26	20.62	-0.7	4.2	1.6	-6.5
Kansas	24.03	25.33	26.41	27.97	28.59	28.21	30.44	31.15	32.56	31.70	30.83	28.91	26.11	25.30	24.80	0.8	4.2	2.6	-6.0
Michigan	25.40	26.79	27.64	28.25	28.64	29.82	30.21	31.47	32.17	33.24	33.14	31.62	29.81	27.96	26.38	1.0	4.4	3.3	-6.8
Minnesota	21.44	21.64	21.18	22.26	22.79	21.82	22.41	22.99	24.05	23.99	22.43	22.11	20.82	19.10	17.56	-3.9	0.4	0.6	-4.9
Missouri	22.04	23.52	24.04	24.84	25.41	26.47	27.01	29.12	29.42	32.35	29.20	28.21	26.85	25.12	23.91	1.9	4.4	2.7	-5.3
Nebraska	22.82	23.01	26.24	24.10	24.31	24.55	24.77	24.66	25.32	25.92	25.96	25.35	24.48	21.91	20.27	-2.5	1.7	1.4	-5.7
North Dakota .	29.91	33.23	32.74	33.55	33.66	33.54	34.09	35.77	37.45	37.96	38.75	36.59	35.04	33.12	31.44	1.5	3.6	5.2	-7.3
Ohio	19.88	21.74	22.31	23.18	23.41	24.68	24.80	26.04	27.40	28.96	29.29	27.60	25.48	25.10	22.36	2.5	4.8	4.6	-6.9
South Dakota .	29.15	31.00	32.59	31.04	32.29	30.77	35.07	35.55	37.39	38.16	37.15	36.62	33.28	32.50	30.61	1.5	1.6	6.4	-6.5
Wisconsin	24.50	25.74	26.93	27.25	28.22	29.11	29.86	30.18	30.79	29.87	29.61	29.16	26.50	28.01	27.46	3.0	4.6	0.5	-2.2
South	20.59	21.84	22.85	23.78	24.29	24.96	25.39	25.97	27.10	27.58	27.17	26.03	24.39	22.54	21.37	0.8	4.4	2.2	-5.8
Alabama	20.96	21.34	22.11	23.32	24.51	25.70	26.49	27.93	29.67	29.88	29.29	29.10	27.70	26.11	24.49	3.5	4.7	3.6	-4.8
Arkansas	17.70	20.54	19.98	21.60	21.97	22.89	23.62	25.14	27.51	28.35	26.58	23.86	23.56	21.56	21.20	3.5	5.2	3.7	-5.4
Delaware	28.10	29.14	31.16	31.84	27.75	28.08	27.85	30.30	29.48	31.05	28.58	26.93	24.35	21.85	20.99	-7.1	-0.0	0.5	-7.6
District of Columbia . . .	17.94	19.51	20.93	21.32	21.54	22.24	22.31	23.49	24.62	23.69	25.06	25.21	21.92	20.82	19.77	1.8	4.3	2.8	-5.3
Florida	16.64	18.74	19.78	21.05	20.49	21.76	22.79	22.12	24.64	26.68	26.62	25.04	23.70	21.87	21.23	4.6	5.1	4.9	-5.4
Georgia	23.33	24.48	25.55	26.53	27.03	27.29	27.56	28.55	29.14	29.42	27.47	27.46	25.68	23.21	23.42	0.1	4.0	0.2	-4.0
Kentucky	19.86	21.68	23.45	23.79	25.20	24.65	25.57	26.38	27.90	29.02	28.24	26.82	25.34	23.09	23.04	3.2	4.8	3.6	-5.2
Louisiana	23.03	24.89	26.01	27.04	28.60	28.80	29.82	31.17	31.30	32.33	32.11	30.26	28.44	25.79	23.94	0.9	5.8	3.3	-8.2
Maryland	18.96	18.58	19.73	20.32	21.29	21.75	21.77	23.32	24.20	24.47	25.05	24.03	23.00	21.81	20.47	1.5	2.8	3.3	-4.6
Mississippi . . .	20.41	21.76	21.41	24.02	24.78	25.65	25.72	26.76	27.68	29.26	29.94	29.02	26.31	25.22	22.51	2.1	5.2	4.3	-7.4
North Carolina	19.17	20.56	21.48	21.29	22.11	22.96	23.14	23.34	24.41	24.48	24.91	23.73	22.18	20.03	20.09	0.9	3.8	1.9	-4.8
Oklahoma	21.79	22.63	23.53	25.23	25.47	27.48	26.36	28.50	29.55	29.61	28.57	27.22	24.41	23.33	23.42	1.6	5.7	1.1	-5.1
South Carolina	18.97	20.85	21.95	21.05	21.41	21.58	20.47	20.44	21.31	22.94	23.45	22.92	21.42	22.98	18.86	-0.1	2.6	1.9	-4.6
Tennessee . . .	21.55	22.83	22.95	24.68	24.67	26.03	26.99	27.54	28.59	29.30	29.04	27.07	25.35	22.75	21.91	0.4	4.5	3.0	-7.1
Texas	22.43	23.60	24.41	25.48	26.45	26.98	27.36	27.37	28.12	27.63	26.96	25.68	24.42	22.22	20.57	-1.9	4.6	-0.0	-6.4
Virginia	20.71	21.27	22.92	24.05	23.67	23.63	24.33	24.77	25.86	26.58	25.69	25.00	23.52	21.09	19.37	-1.3	2.9	2.1	-6.3
West Virginia .	19.61	21.41	25.12	25.05	25.91	26.80	27.98	29.49	29.95	28.69	28.52	27.06	23.43	23.73	21.61	2.0	7.2	1.7	-6.9
West	22.60	23.52	24.09	25.18	26.30	26.06	26.13	27.33	27.13	27.92	28.06	26.89	25.68	23.41	22.16	-0.4	3.5	2.0	-5.9
Alaska	29.47	27.66	31.93	27.57	36.75	32.26	31.59	36.21	29.34	31.10	35.24	30.29	27.62	24.23	20.71	-8.8	2.8	3.0	-14.5
Arizona	24.52	26.52	25.86	25.72	27.79	26.41	28.20	29.58	31.02	32.09	32.12	29.94	27.52	22.40	20.34	-4.2	1.9	5.7	-11.8
California	21.34	21.89	22.75	24.00	25.23	25.27	25.51	27.00	26.73	27.32	27.56	26.97	25.89	23.60	22.77	1.4	3.9	2.3	-4.8
Colorado	25.29	26.53	26.02	27.96	30.27	28.96	29.08	30.85	31.12	31.60	31.61	28.58	27.32	26.54	25.01	-0.3	3.7	2.6	-6.6
Hawaii	14.88	15.57	18.22	19.07	20.72	19.87	20.87	18.28	19.82	18.97	20.74	20.32	19.47	17.36	16.94	2.1	5.0	0.9	-3.8
Idaho	24.38	28.50	27.04	29.44	32.13	29.43	29.55	30.19	30.65	29.81	28.84	26.99	24.42	22.77	22.97	-1.4	5.1	-0.6	-5.9

Table 4-3.—Continued

Region and state	1975-76	1976-77	1977-78	1978-79	1979-80	1980-81	1981-82	1982-83	1983-84	1984-85	1985-86	1986-87	1987-88	1988-89	1989-90	Change in percentage points: 1975-76 to 1989-90	1975-76 to 1980-81	1980-81 to 1985-86	1985-86 to 1989-90
West Continued																			
Montana	33.42	34.92	36.48	37.51	35.15	36.28	37.70	37.76	38.98	39.59	36.71	36.01	33.81	30.10	29.85	-3.6	2.9	0.4	-6.9
Nevada	23.56	23.72	23.72	24.08	27.69	25.25	23.69	24.35	22.83	22.99	23.88	23.64	21.41	20.07	18.39	-5.2	1.7	-1.4	-5.5
New Mexico ..	22.65	24.34	26.58	28.16	29.26	28.51	28.78	31.05	31.05	32.58	33.04	32.85	30.66	26.60	25.17	2.5	5.9	4.5	-7.9
Oregon	21.86	24.17	24.10	24.13	23.74	24.63	23.91	24.50	22.66	23.85	26.07	24.02	23.44	20.98	19.76	-2.1	2.8	1.4	-6.3
Utah.........	22.62	23.93	23.11	24.67	25.04	25.48	24.38	25.96	25.81	27.13	27.77	25.34	24.12	22.97	20.56	-2.1	2.9	2.3	-7.2
Washington...	23.45	23.66	24.68	25.08	25.31	24.68	23.75	24.24	23.41	24.67	23.41	22.88	22.37	21.22	19.05	-4.4	1.2	-1.3	-4.4
Wyoming	37.94	38.97	40.47	40.34	39.56	41.21	40.51	35.63	38.74	37.75	35.37	34.40	32.07	29.20	26.12	-11.8	3.3	-5.8	-9.2
U.S. Service Schools......	35.24	40.77	38.16	41.84	42.46	40.99	42.80	39.02	43.65	46.45	42.89	46.97	36.88	36.20	34.61	-0.6	5.7	1.9	-8.3

SOURCE: U.S. Department of Education, National Center for Education Statistics, HEGIS, "Degrees and Other Formal Awards Conferred" surveys; and IPEDS,"Completions" surveys.

Table 4-4.—Master's degrees conferred in science, by region and state:1975-76 to 1989-90

Region and state	1975-76	1976-77	1977-78	1978-79	1979-80	1980-81	1981-82	1982-83	1983-84	1984-85	1985-86	1986-87	1987-88	1988-89	1989-90	Percent change 1975-76 to 1989-90	1975-76 to 1980-81	1980-81 to 1985-86	1985-86 to 1989-90
United States . . .	46,218	47,535	49,445	49,607	50,663	52,196	54,367	56,979	59,454	60,824	63,071	63,662	65,246	67,208	68,526	48.3	12.9	20.8	8.6
Northeast	12,266	12,027	12,804	12,927	13,335	13,459	14,382	15,261	16,004	16,337	16,991	17,943	17,405	18,758	18,640	52.0	9.7	26.2	9.7
Connecticut . .	751	833	690	670	754	918	917	947	1,037	1,085	1,264	1,124	1,107	1,146	1,176	56.6	22.2	37.7	-7.0
Maine	64	74	76	92	78	69	57	76	64	83	92	95	80	93	115	79.7	7.8	33.3	25.0
Massachusetts	2,526	2,497	2,580	2,646	2,882	2,779	2,887	3,088	3,020	3,136	3,451	3,680	3,750	3,947	3,882	53.7	10.0	24.2	12.5
New Hampshire	120	115	98	108	107	122	143	185	181	174	175	202	173	194	267	122.5	1.7	43.4	52.6
New Jersey . . .	1,395	1,246	1,257	1,261	1,313	1,348	1,518	1,447	1,589	1,578	1,728	1,815	1,966	2,336	2,284	63.7	-3.4	28.2	32.2
New York	4,973	4,953	5,624	5,508	5,516	5,599	5,882	6,166	6,511	6,743	6,646	7,215	6,558	7,087	7,141	43.6	12.6	18.7	7.4
Pennsylvania .	2,095	2,003	2,164	2,310	2,356	2,295	2,618	2,911	3,123	3,093	3,209	3,374	3,310	3,491	3,372	61.0	9.5	39.8	5.1
Rhode Island .	226	222	191	199	186	202	218	273	320	307	282	317	331	317	298	31.9	-10.6	39.6	5.7
Vermont	116	84	124	133	143	127	142	168	159	138	144	121	130	147	105	-9.5	9.5	13.4	-27.1
Midwest	11,831	12,373	12,853	12,780	12,949	13,132	13,774	14,067	14,524	14,855	15,322	14,969	15,238	15,602	16,157	36.6	11.0	16.7	5.4
Illinois	2,413	2,447	2,552	2,569	2,605	2,713	2,918	2,857	3,079	3,240	3,563	3,371	3,311	3,525	3,568	47.9	12.4	31.3	0.1
Indiana	1,106	1,190	1,160	1,109	1,143	1,174	1,211	1,282	1,339	1,329	1,261	1,230	1,275	1,265	1,303	17.8	6.1	7.4	3.3
Iowa	557	608	561	548	554	563	553	539	653	642	607	609	679	600	709	27.3	1.1	7.8	16.8
Kansas	500	525	678	570	524	567	592	605	634	674	708	697	675	715	721	44.2	13.4	24.9	1.8
Michigan	2,354	2,197	2,296	2,364	2,437	2,486	2,447	2,435	2,466	2,486	2,592	2,428	2,534	2,651	2,687	14.1	5.6	4.3	3.7
Minnesota	663	735	797	710	723	699	743	824	777	851	837	904	882	912	992	49.6	5.4	19.7	18.5
Missouri	964	1,162	1,245	1,161	1,177	1,180	1,263	1,393	1,348	1,441	1,340	1,277	1,321	1,308	1,482	53.7	22.4	13.6	10.6
Nebraska	223	226	299	283	277	267	269	313	302	295	314	287	302	292	262	17.5	19.7	17.6	-16.6
North Dakota .	117	143	131	144	125	142	145	167	207	202	197	217	219	198	218	86.3	21.4	38.7	10.7
Ohio	1,832	1,958	1,885	2,000	2,100	2,124	2,220	2,299	2,316	2,307	2,485	2,558	2,618	2,707	2,758	50.5	15.9	17.0	11.0
South Dakota .	133	147	134	168	117	127	149	163	179	197	190	151	175	165	185	39.1	-4.5	49.6	-2.6
Wisconsin	969	1,035	1,115	1,154	1,167	1,090	1,264	1,190	1,224	1,191	1,228	1,240	1,247	1,264	1,272	31.3	12.5	12.7	3.6
South	12,083	12,626	13,072	13,136	13,570	14,119	14,172	15,101	15,975	16,524	17,563	17,522	18,652	18,773	18,769	55.3	16.9	24.4	6.9
Alabama	408	460	497	585	678	651	743	764	820	860	935	930	972	1,054	1,117	173.8	59.6	43.6	19.5
Arkansas	319	341	331	359	314	352	393	443	448	431	461	432	406	281	307	-3.8	10.3	31.0	-33.4
Delaware	129	102	129	118	114	117	148	151	149	134	145	144	136	150	149	15.5	-9.3	23.9	2.8
District of Columbia . . .	995	1,096	1,076	964	1,081	1,102	991	1,222	1,035	1,093	1,068	1,048	1,146	1,150	1,042	4.7	10.8	-3.1	-2.4
Florida	945	912	895	915	989	966	1,061	1,161	1,222	1,306	1,488	1,590	1,782	1,906	1,980	109.5	2.2	54.0	33.1
Georgia	972	989	999	943	970	1,053	1,024	965	1,149	1,102	1,149	1,187	1,277	1,274	1,259	29.5	8.3	9.1	9.6
Kentucky	415	457	445	473	497	513	500	568	583	519	603	575	594	637	618	48.9	23.6	17.5	2.5
Louisiana	590	612	641	617	686	694	705	761	841	929	965	941	978	895	892	51.2	17.6	39.0	-7.6
Maryland	824	813	875	906	890	997	958	1,014	1,187	1,240	1,314	1,404	1,506	1,561	1,772	115.0	21.0	31.8	34.9
Mississippi . . .	315	368	405	334	368	410	377	362	378	509	508	497	513	501	466	47.9	30.2	23.9	-8.3
North Carolina	827	866	1,002	959	991	1,026	1,056	1,086	1,180	1,200	1,234	1,269	1,302	1,265	1,251	51.3	24.1	20.3	1.4
Oklahoma	652	588	542	653	535	575	513	530	536	716	749	655	704	740	689	5.7	-11.8	30.3	-8.0
South Carolina	339	388	434	402	435	454	501	478	534	535	561	578	608	578	669	97.3	33.9	23.6	19.3
Tennessee . . .	764	845	785	770	815	793	792	755	712	771	788	769	857	911	860	12.6	3.8	-0.6	9.1
Texas	2,578	2,686	2,858	2,998	3,048	3,103	3,033	3,311	3,549	3,632	3,993	3,868	4,026	3,970	3,803	47.5	20.4	28.7	-4.8
Virginia	735	855	854	849	862	1,012	1,024	1,133	1,267	1,197	1,258	1,313	1,498	1,591	1,619	120.3	37.7	24.3	28.7
West Virginia .	276	248	304	291	297	301	353	397	385	350	344	322	347	309	276	0.0	9.1	14.3	-19.8
West	9,587	10,021	10,224	10,388	10,366	11,002	11,528	12,075	12,386	12,591	12,591	12,829	13,555	13,634	14,650	52.8	14.8	14.4	16.4
Alaska	29	46	30	39	57	63	55	87	76	100	101	92	107	83	97	234.5	117.2	60.3	-4.0
Arizona	531	581	553	607	568	596	753	755	773	808	816	853	930	875	964	81.5	12.2	36.9	18.1
California	5,731	5,698	5,862	5,910	5,871	6,294	6,531	6,961	7,357	7,271	7,225	7,351	7,691	7,797	8,498	48.3	9.8	14.8	17.6
Colorado	698	798	876	826	796	882	878	989	933	911	949	972	1,191	1,127	1,213	73.8	26.4	7.6	27.8
Hawaii	288	308	306	335	337	336	312	337	281	318	196	283	258	231	234	-18.8	16.7	-41.7	19.4

Table 4-4.—Continued

Region and state	1975-76	1976-77	1977-78	1978-79	1979-80	1980-81	1981-82	1982-83	1983-84	1984-85	1985-86	1986-87	1987-88	1988-89	1989-90	Percent change: 1975-76 to 1989-90	Percent change: 1975-76 to 1980-81	Percent change: 1980-81 to 1985-86	Percent change: 1985-86 to 1989-90
West Continued																			
Idaho	148	177	173	176	188	186	163	192	171	197	216	208	174	181	222	50.0	25.7	16.1	2.8
Montana	169	171	149	143	144	158	168	201	185	212	222	220	217	208	208	23.1	-6.5	40.5	-6.3
Nevada	70	77	70	53	70	83	75	59	70	75	113	93	98	123	136	94.3	18.6	36.1	20.4
New Mexico	295	286	314	341	328	367	339	325	383	378	383	359	376	470	415	40.7	24.4	4.4	8.4
Oregon	335	413	468	402	447	478	579	509	542	531	532	524	579	638	651	94.3	42.7	11.3	22.4
Utah	441	485	474	481	536	545	523	507	559	639	574	644	743	678	635	44.0	23.6	5.3	10.6
Washington	752	863	825	955	934	912	1,064	1,029	916	1,017	1,123	1,100	1,078	1,090	1,271	69.0	21.3	23.1	13.2
Wyoming	100	118	124	120	90	102	88	124	140	134	141	130	113	133	106	6.0	2.0	38.2	-24.8
U.S. Service Schools	451	488	492	376	443	484	511	475	565	517	604	399	396	441	310	-31.3	7.3	24.8	-48.7
Outlying Areas	255	212	242	241	166	244	222	231	367	225	233	280	221	231	231	-9.4	-4.3	-4.5	-0.9
American Samoa	0	0	0	0	0	0	0	0	0	0	0	0	0	0	0	([1])	([1])	([1])	([1])
Guam	3	5	3	1	9	3	5	6	2	2	4	3	2	3	2	-33.3	0.0	33.3	-50.0
Northern Marianas	0	0	0	0	0	0	0	0	0	0	0	0	0	0	0	([1])	([1])	([1])	([1])
Puerto Rico	252	207	239	240	157	241	217	225	365	223	229	277	219	228	229	-9.1	-4.4	-5.0	0.0
Trust Territories	0	0	0	0	0	0	0	0	0	0	0	0	0	0	0	([1])	([1])	([1])	([1])
Virgin Islands	0	0	0	0	0	0	0	0	0	0	0	0	0	0	0	([1])	([1])	([1])	([1])

[1] Insufficient data for calculating a percent change.

—Data not available.

SOURCE: U.S. Department of Education, National Center for Education Statistics, HEGIS, "Degrees and Other Formal Awards Conferred" surveys; and IPEDS,"Completions" surveys.

Table 4-5.—Doctor's degrees conferred in science, by region and state: 1975-76 to 1989-90

Region and state	1975-76	1976-77	1977-78	1978-79	1979-80	1980-81	1981-82	1982-83	1983-84	1984-85	1985-86	1986-87	1987-88	1988-89	1989-90	Percent change			
																1975-76 to 1989-90	1975-76 to 1980-81	1980-81 to 1985-86	1985-86 to 1989-90
United States	11,393	10,971	10,687	11,041	11,234	11,566	11,905	12,007	12,310	12,725	13,062	13,546	14,460	15,071	16,415	44.1	1.5	12.9	25.7
Northeast	3,005	2,873	2,816	2,922	2,889	2,974	3,106	3,000	3,146	3,176	3,285	3,410	3,560	3,804	4,092	36.2	-1.0	10.5	24.6
Connecticut	155	171	161	148	160	175	159	188	221	185	205	238	234	210	252	62.6	12.9	17.1	22.9
Maine	12	17	17	7	6	10	11	9	6	10	16	15	12	13	22	83.3	-16.7	60.0	37.5
Massachusetts	735	668	658	680	687	724	713	697	731	760	810	800	873	892	981	33.5	-1.5	11.9	21.1
New Hampshire	49	38	33	36	44	48	43	45	38	33	35	46	41	35	52	6.1	-2.0	-27.1	48.6
New Jersey	223	272	242	245	216	259	256	275	255	283	270	275	359	303	364	63.2	16.1	4.2	34.8
New York	1,125	1,072	1,072	1,144	1,116	1,132	1,198	1,111	1,152	1,169	1,220	1,298	1,236	1,427	1,518	34.9	0.6	7.8	24.4
Pennsylvania	598	532	535	558	547	537	618	571	628	630	617	625	673	781	776	29.8	-10.2	14.9	25.8
Rhode Island	86	87	80	85	97	71	92	88	98	90	91	93	108	115	102	18.6	-17.4	28.2	12.1
Vermont	22	16	18	19	16	18	16	16	17	16	21	20	24	28	25	13.6	-18.2	16.7	19.0
Midwest	3,179	2,987	2,911	3,140	3,136	3,212	3,295	3,147	3,278	3,513	3,443	3,687	3,849	3,946	4,223	32.8	1.0	7.2	22.7
Illinois	625	644	582	652	670	723	670	650	653	727	731	776	841	816	938	50.1	15.7	1.1	28.3
Indiana	398	375	354	364	359	369	359	354	402	389	390	404	413	444	477	19.8	-7.3	5.7	22.3
Iowa	265	193	201	238	222	242	266	215	230	266	253	323	317	269	290	9.4	-8.7	4.5	14.6
Kansas	138	127	168	145	143	147	136	145	129	182	125	146	151	186	170	23.2	6.5	-15.0	36.0
Michigan	434	438	422	464	454	426	500	500	484	469	483	510	539	571	619	42.6	-1.8	13.4	28.2
Minnesota	193	180	193	187	205	245	205	200	247	266	287	245	284	282	328	69.9	26.9	17.1	14.3
Missouri	158	224	159	196	196	180	192	152	181	188	185	179	197	231	261	65.2	13.9	2.8	41.1
Nebraska	58	59	51	67	78	95	84	108	80	80	72	86	91	86	93	60.3	63.8	-24.2	29.2
North Dakota	30	34	19	42	41	32	27	28	24	27	43	39	51	46	38	26.7	6.7	34.4	-11.6
Ohio	500	406	424	420	434	442	476	464	513	534	505	582	531	620	612	22.4	-11.6	14.3	21.2
South Dakota	8	10	10	6	7	5	7	3	7	11	11	15	14	16	14	75.0	-37.5	120.0	27.3
Wisconsin	372	297	328	359	327	306	373	328	328	374	358	382	420	379	383	3.0	-17.7	17.0	7.0
South	2,643	2,596	2,533	2,653	2,624	2,634	2,765	2,947	2,959	3,102	3,274	3,421	3,728	3,911	4,311	63.1	-0.3	24.3	31.7
Alabama	54	78	69	68	86	73	84	94	104	103	100	126	125	172	166	207.4	35.2	37.0	66.0
Arkansas	43	24	25	25	32	44	45	179	59	49	69	50	40	36	62	44.2	2.3	56.8	-10.1
Delaware	36	29	43	25	36	29	64	38	0	46	50	50	64	61	61	69.4	-19.4	72.4	22.0
District of Columbia	122	118	138	133	120	134	133	116	110	118	144	113	131	130	142	16.4	9.8	7.5	-1.4
Florida	239	266	208	249	231	201	187	215	217	249	238	273	338	336	391	63.6	-15.9	18.4	64.3
Georgia	148	166	179	146	136	171	192	183	181	214	208	202	234	299	266	79.7	15.5	21.6	27.9
Kentucky	58	57	53	63	78	76	58	68	81	76	65	90	86	104	92	58.6	31.0	-14.5	41.5
Louisiana	123	101	75	115	98	81	98	97	97	101	135	134	157	172	217	76.4	-34.1	66.7	60.7
Maryland	251	261	270	252	222	232	244	258	267	306	306	326	347	354	427	70.1	-7.6	31.9	39.5
Mississippi	84	73	56	47	58	71	81	61	73	59	68	76	74	75	88	4.8	-15.5	-4.2	29.4
North Carolina	279	256	293	297	303	295	349	370	389	334	362	393	412	363	457	63.8	5.7	22.7	26.2
Oklahoma	172	145	135	168	151	124	135	128	128	139	158	135	150	133	166	-3.5	-27.9	27.4	5.1
South Carolina	77	66	68	94	76	72	77	87	87	99	95	118	138	118	136	76.6	-6.5	31.9	43.2
Tennessee	149	161	130	126	145	132	117	145	163	160	141	136	150	150	185	24.2	-11.4	6.8	31.2
Texas	575	571	565	598	612	633	630	649	700	734	793	875	885	981	1,035	80.0	10.1	25.3	30.5
Virginia	186	179	177	202	188	228	221	218	263	273	296	290	347	373	380	104.3	22.6	29.8	28.4
West Virginia	47	45	49	45	52	38	50	41	40	42	46	34	50	54	40	-14.9	-19.1	21.1	-13.0
West	2,538	2,501	2,420	2,311	2,570	2,735	2,724	2,900	2,908	2,912	3,041	3,010	3,323	3,410	3,789	49.3	7.8	11.2	24.6
Alaska	8	3	2	5	0	2	1	5	5	8	7	3	11	9	5	-37.5	-75.0	250.0	-28.6
Arizona	142	164	146	125	168	139	185	167	172	192	192	205	244	288	271	90.8	-2.1	38.1	41.1
California	1,498	1,446	1,421	1,298	1,479	1,632	1,594	1,661	1,697	1,666	1,777	1,699	1,853	1,901	2,205	47.2	8.9	8.9	24.1
Colorado	233	201	193	209	214	225	216	248	235	248	253	264	292	290	337	44.6	-3.4	12.4	33.2
Hawaii	44	58	76	44	45	63	57	51	49	62	53	67	61	79	46	4.5	43.2	-15.9	-13.2
Idaho	31	29	32	34	25	34	33	33	30	27	20	29	38	39	53	71.0	9.7	-41.2	165.0

Table 4-5.—Continued

Region and state	1975-76	1976-77	1977-78	1978-79	1979-80	1980-81	1981-82	1982-83	1983-84	1984-85	1985-86	1986-87	1987-88	1988-89	1989-90	Percent change 1975-76 to 1989-90	Percent change 1975-76 to 1980-81	Percent change 1980-81 to 1985-86	Percent change 1985-86 to 1989-90
West Continued																			
Montana	32	25	22	19	20	14	22	33	22	25	34	27	30	25	35	9.4	-56.3	142.9	2.9
Nevada	8	16	10	8	11	7	11	8	7	6	8	7	11	13	20	150.0	-12.5	14.3	150.0
New Mexico	46	46	51	55	55	55	46	76	60	77	92	82	98	108	112	143.5	19.6	67.3	21.7
Oregon	124	154	123	132	120	163	132	206	197	194	182	140	160	158	187	50.8	31.5	11.7	2.7
Utah	127	122	125	123	148	130	147	133	156	132	126	163	181	172	174	37.0	2.4	-3.1	38.1
Washington	226	204	193	241	250	239	253	253	254	250	260	295	307	297	310	37.2	5.8	8.8	19.2
Wyoming	19	33	26	18	35	32	27	26	24	25	37	29	37	31	34	78.9	68.4	15.6	-8.1
U.S. Service Schools	28	14	7	15	15	11	15	13	19	22	19	18	0	0	0	-100.0	-60.7	72.7	-100.0
Outlying Areas	7	9	8	8	8	10	9	8	10	8	9	7	25	12	14	100.0	42.9	-10.0	55.6
American Samoa	0	0	0	0	0	0	0	0	0	0	0	0	0	0	0	([1])	([1])	([1])	([1])
Guam	0	0	0	0	0	0	0	0	0	0	0	0	0	0	0	([1])	([1])	([1])	([1])
Northern Marianas	0	0	0	0	0	0	0	0	0	0	0	0	0	0	0	([1])	([1])	([1])	([1])
Puerto Rico	7	9	8	8	8	10	9	8	10	8	9	7	25	12	14	100.0	42.9	-10.0	55.6
Trust Territories	0	0	0	0	0	0	0	0	0	0	0	0	0	0	0	([1])	([1])	([1])	([1])
Virgin Islands	0	0	0	0	0	0	0	0	0	0	0	0	0	0	0	([1])	([1])	([1])	([1])

[1]Insufficient data for calculating a percent change.

—Data not available.

SOURCE: U.S. Department of Education, National Center for Education Statistics, HEGIS, "Degrees and Other Formal Awards Conferred" surveys; and IPEDS,"Completions" surveys.

THIS PAGE INTENTIONALLY LEFT BLANK

5. Supply of Graduates in Science: Associate, Bachelor's, Master's and Doctor's Degrees, by Science Field, Region, and State

In earlier chapters, data on science and mathematics degrees were presented divided by degree level, region, and state. In this chapter, the data on science degrees are further divided into six science fields: agricultural sciences, computer sciences, engineering, health sciences, life sciences, and physical sciences.

In recent years the numbers of bachelor's degrees attained in science have dropped. Yet, these declining numbers are not represented in every field in all regions and states. Also, some fields are losing popularity at the bachelor's level, while gaining in the numbers of graduate degrees.

Associate Degrees

Regions

Between 1982–83[1] and 1989–90, associate degrees declined in almost every region in the country in almost every science field (text table 3). The exceptions were associates awarded in life sciences, which rose in the South and West, associates awarded in physical sciences which rose in the West, and associates awarded in health sciences which rose in the South. During that period associate degrees in agricultural sciences declined the greatest in the Midwest (38 percent) and in the West (44 percent). Computer sciences associate degrees, which declined nationally by over 21 percent from 1982–83 to 1989–90, also had significant decreases in the Midwest (42 percent) and in the West (29 percent). In contrast, associates in engineering, down 11 percent nationally during that same period, declined the most in the Northeast (22 percent) (text table 3).

States

Tables 5–1 through 5–6 present state-by-state data on associate degrees by science field. Between 1982–83 and 1989–90 the numbers of associate degrees awarded in agricultural sciences dropped in almost every state in the country; the exceptions were Rhode Island, Vermont, Alabama, Mississippi, Oklahoma, Tennessee, Utah, and Wyoming. Other fields fared only slightly better. Associates in computer sciences declined in 31 states and the District of Columbia, as did associates in engineering in 34 states and the District of Columbia during that same period. Also, between 1982–83 and 1989–90, numbers of associate degrees conferred in health sciences declined in 26 states and the District of Columbia.

Some states had increases in life sciences and physical sciences between 1982–83 and 1989–90, but in most cases the numbers involved were relatively small. In both fields, many states did not report enough data for trends to be analyzed.

Bachelor's Degrees

Regions

As with associate degrees, numbers of bachelor's degrees in agricultural sciences declined sharply. Between 1975–76 and 1989–90, bachelor's degrees conferred in agricultural, life, and physical sciences dropped in every region in the country (text table 4). The Northeast had a decline of 40 percent in bachelor's degrees conferred in life sciences during that period, the largest of all the regions, while the smallest decrease was in the West (24 percent). Between 1975–76 and 1989–90 bachelor's degrees conferred in health sciences increased in every region except the West. Other increases were seen in engineering and computer sciences, as bachelor's degrees in these fields grew in number in every region during this period.

The West had the largest increase in bachelor's degrees in computer sciences between 1975–76 and 1989–90, 490 percent, from only 841 in 1975–76 to 4,966 in 1989–90. The increases in the other regions were also considerable, between 340 and nearly 400 percent in the South, Northeast, and Midwest. As with computer sciences, increases in numbers of bachelor's in engineering were largest in the West (95 percent), while the other regions had increases of between 57 and 87 percent.

The period of 1985–86 to 1989–90 was not a fruitful one for the production of bachelor's degrees in the science fields. Every region had a decline in the number of bachelor's degrees awarded in every science field between 1985–86 and 1989–90, with the exception of life sciences, where the South and West had increases of 2 and 5 percent, respectively. The South or the Midwest was a leader in the awarding of bachelor's degrees in every field, and the West was

[1] The academic year 1982–83 is the first year for which the field of study data are consistent for associate degrees.

Text table 3.—Associate degrees by science field and region: 1982-83 to 1989-90

Field and region	1982-83	1985-86	1989-90	Percent change: 1982-83 to 1989-90	Percent change: 1982-83 to 1985-86	Percent change: 1985-86 to 1989-90
Agricultural sciences						
United States	7,645	5,741	4,832	-36.8	-24.9	-15.8
Northeast	1,742	1,226	1,149	-34.0	-29.6	-6.3
Midwest	2,782	2,087	1,715	-38.4	-25.0	-17.8
South	1,501	1,208	1,065	-29.0	-19.5	-11.8
West	1,620	1,220	903	-44.3	-24.7	-26.0
U.S.S.S.	—	—	—	(1)	(1)	(1)
Engineering						
United States	60,629	63,339	54,131	-10.7	4.5	-14.5
Northeast	14,566	13,250	11,367	-22.0	-9.0	-14.2
Midwest	16,588	17,334	14,834	-10.6	4.5	-14.4
South	16,116	16,743	14,445	-10.4	3.9	-13.7
West	11,163	12,257	10,261	-8.1	9.8	-16.3
U.S.S.S.	2,196	3,755	3,224	46.8	71.0	-14.1
Life sciences						
United States	981	998	1,034	5.4	1.7	3.6
Northeast	101	104	86	-14.9	3.0	-17.3
Midwest	233	180	100	-57.1	-22.7	-44.4
South	143	235	173	21.0	64.3	-26.4
West	504	479	675	33.9	-5.0	40.9
U.S.S.S.	0	0	—	(1)	(1)	(1)
Computer sciences						
United States	9,670	10,704	7,604	-21.4	10.7	-29.0
Northeast	1,928	1,735	1,735	-10.0	-10.0	0.0
Midwest	3,260	2,896	1,881	-42.3	-11.2	-35.0
South	2,207	3,088	2,118	-4.0	39.9	-31.4
West	2,079	2,707	1,483	-28.7	30.2	-45.2
U.S.S.S.	196	278	387	97.4	41.8	39.2
Health sciences						
United States	65,749	66,559	64,128	-2.5	1.2	-3.7
Northeast	14,495	14,824	14,141	-2.4	2.3	-4.6
Midwest	17,804	18,879	17,381	-2.4	6.0	-7.9
South	21,869	21,502	21,892	0.1	-1.7	1.8
West	11,210	10,905	10,227	-8.8	-2.7	-6.2
U.S.S.S.	371	449	487	31.3	21.0	8.5
Physical sciences						
United States	3,103	2,107	2,135	-31.2	-32.1	1.3
Northeast	721	543	362	-49.8	-24.7	-33.3
Midwest	746	440	456	-38.9	-41.0	3.6
South	859	481	439	-48.9	-44.0	-8.7
West	600	556	692	15.3	-7.3	24.5
U.S.S.S.	177	87	186	5.1	-50.8	113.8
Total science						
United States	147,777	149,448	133,864	-9.4	1.1	-10.4
Northeast	33,553	31,682	28,840	-14.0	-5.6	-9.0
Midwest	41,413	41,816	36,367	-12.2	1.0	-13.0
South	42,695	43,257	40,132	-6.0	1.3	-7.2
West	27,176	28,124	24,241	-10.8	3.5	-13.8
U.S.S.S.	2,940	4,569	4,284	45.7	55.4	-6.2

[1] Insufficient data for calculating a percent change.

—Data not available.

Note: U.S.S.S.=U. S. Service Schools.

SOURCE: U.S. Department of Education, National Center for Education Statistics, HEGIS, "Degrees and Other Formal Awards Conferred" surveys; and IPEDS, "Completions" surveys.

Text table 4.—Bachelor's degrees by science field and region: 1975-76 to 1989-90

Field and region	1975-76	1980-81	1985-86	1989-90	Percent change 1975-76 to 1989-90	Percent change 1975-76 to 1980-81	Percent change 1980-81 to 1985-86	Percent change 1985-86 to 1989-90
Agricultural sciences								
United States	19,402	21,886	16,823	13,070	-32.6	12.8	-23.1	-22.3
Northeast	2,796	3,000	2,299	1,980	-29.2	7.3	-23.4	-13.9
Midwest	6,208	7,238	5,616	4,343	-30.0	16.6	-22.4	-22.7
South	5,708	6,645	5,111	4,060	-28.9	16.4	-23.1	-20.6
West	4,690	5,003	3,797	2,687	-42.7	6.7	-24.1	-29.2
U.S.S.S.	—	—	—	—	([1])	([1])	([1])	([1])
Engineering								
United States	46,331	75,000	95,953	82,110	77.2	61.9	27.9	-14.4
Northeast	11,787	17,442	22,070	18,487	56.8	48.0	26.5	-16.2
Midwest	13,195	21,570	27,658	24,617	86.6	63.5	28.2	-11.0
South	12,552	21,795	27,395	22,383	78.3	73.6	25.7	-18.3
West	8,098	13,285	17,775	15,829	95.5	64.1	33.8	-10.9
U.S.S.S.	699	908	1,055	794	13.6	29.9	16.2	-24.7
Life sciences								
United States	54,275	43,216	38,524	37,170	-31.5	-20.4	-10.9	-3.5
Northeast	14,909	12,317	10,480	8,939	-40.0	-17.4	-14.9	-14.7
Midwest	13,884	10,888	9,839	9,394	-32.3	-21.6	-9.6	-4.5
South	14,520	11,720	10,308	10,546	-27.4	-19.3	-12.0	2.3
West	10,878	8,241	7,868	8,256	-24.1	-24.2	-4.5	4.9
U.S.S.S.	84	50	29	35	-58.3	-40.5	-42.0	20.7
Total science								
United States	200,938	242,523	279,455	234,731	16.8	20.7	15.2	-16.0
Northeast	50,075	60,020	68,281	54,749	9.3	19.9	13.8	-19.8
Midwest	56,859	69,292	79,850	69,198	21.7	21.9	15.2	-13.3
South	56,373	69,974	80,758	67,224	19.2	24.1	15.4	-16.8
West	36,524	41,897	48,983	42,423	16.2	14.7	16.9	-13.4
U.S.S.S.	1,107	1,340	1,583	1,137	2.7	21.0	18.1	-28.2
Computer sciences								
United States	5,652	15,121	41,889	27,434	385.4	167.5	177.0	-34.5
Northeast	1,409	4,303	11,735	6,767	380.3	205.4	172.7	-42.3
Midwest	1,642	4,301	10,760	7,237	340.7	161.9	150.2	-32.7
South	1,701	4,383	12,823	8,381	392.7	157.7	192.6	-34.6
West	841	2,094	6,430	4,966	490.5	149.0	207.1	-22.8
U.S.S.S.	59	40	141	83	40.7	-32.2	252.5	-41.1
Health sciences								
United States	53,813	63,348	64,535	58,816	9.3	17.7	1.9	-8.9
Northeast	13,232	16,722	16,492	14,567	10.1	26.4	-1.4	-11.7
Midwest	16,295	19,503	20,266	19,232	18.0	19.7	3.9	-5.1
South	16,025	17,851	18,894	17,314	8.0	11.4	5.8	-8.4
West	8,261	9,272	8,883	7,703	-6.8	12.2	-4.2	-13.3
U.S.S.S.	—	—	0	—	([1])	([1])	([1])	([1])
Physical sciences								
United States	21,465	23,952	21,731	16,131	-24.8	11.6	-9.3	-25.8
Northeast	5,942	6,236	5,205	4,009	-32.5	4.9	-16.5	-23.0
Midwest	5,635	5,792	5,711	4,375	-22.4	2.8	-1.4	-23.4
South	5,867	7,580	6,227	4,540	-22.6	29.2	-17.8	-27.1
West	3,756	4,002	4,230	2,982	-20.6	6.5	5.7	-29.5
U.S.S.S.	265	342	358	225	-15.1	29.1	4.7	-37.2

[1] Insufficient data for calculating a percent change.

—Data not available.

Note: U.S.S.S.=U. S. Service Schools.

SOURCE: U.S. Department of Education, National Center for Education Statistics, HEGIS, "Degrees and Other Formal Awards Conferred" surveys; and IPEDS, "Completions" surveys.

usually last. The exception was in agricultural sciences, a field in which the Northeast was the region with the least number of bachelor's degrees conferred.

States

Tables 5–7 through 5–12 present state-by-state data on bachelor's degrees by field. Only four states had increases in agricultural sciences between 1975–76 and 1989–90: Alaska (45 percent), New York (24 percent), Wisconsin (17 percent), and Delaware (9 percent). With the exception of a small increase in Nevada (3 percent), there were decreases in the numbers of bachelor's degrees conferred in life sciences between 1975–76 and 1989–90 in every state in the Nation and the District of Columbia. These decreases ranged from 11 percent in North Carolina to 55 percent in Wyoming. Physical sciences had a similar pattern. Between 1975–76 and 1989–90, bachelor's degrees in physical sciences declined in all but five states.

Twenty-nine states had increases in bachelor's degrees conferred in health sciences between 1975–76 and 1989–90. The largest were in Maine (294 percent) and Arkansas (97 percent). The largest decline in the number of bachelor's degrees in this field was in the state of Washington (38 percent). The awarding of bachelor's degrees in computer sciences exploded across most of the country between 1975–76 and 1989–90. With the exception of Mississippi, every state that reported bachelor's degrees conferred in computer sciences in 1975–76 had increases of over 100 percent between 1975–76 and 1989–90. The District of Columbia had a 283 percent increase during that time.

Every state and the District of Columbia had increases in bachelor's degrees in engineering from 1975–76 to 1989–90, ranging from 13 percent in Delaware to 188 percent in Florida. It is interesting to note increases of over 100 percent in a number of states with small populations, such as New Hampshire, North Dakota, and Alaska, along with such increases in the larger states of Alabama, Florida, Maryland, Wisconsin, California, and Ohio.

While there were increases in some science areas, such as engineering, between 1975–76 to 1989–90, there was a dramatic shift during the later part of this period. Between 1985–86 and 1989–90 a decrease in bachelor's degrees conferred in science was reflected in almost all states in practically every field. In physical sciences, the numbers of bachelor's degrees awarded between 1985–86 and 1989–90 decreased in every state and the District of Columbia. Other science fields showed similar patterns in numbers of bachelor's degrees conferred: only two states had increases in agricultural sciences, three states increased or were stable in engineering, four states increased or were stable in computer sciences, and only nine states increased in health sciences. There were increases in 17 states in bachelor's degrees in life sciences. However, most of these increases were in single digits.

Master's Degrees

Regions

Master's degrees in computer sciences, engineering, and health sciences increased in every region between 1975–76 and 1989–90 (text table 5). In contrast, in life sciences there were declines in all the regions as well as at the national level. The largest were in the Midwest and South. Agricultural sciences and physical sciences saw little change at the national level, but both had regional increases and decreases. The South and West had small declines in agricultural sciences, while the Northeast and Midwest increased. In physical sciences, the South and West had increases, the Midwest had a minor decrease, and the Northeast had a more substantial (22 percent) decline.

There were substantial regional differences in the rate of growth for master's degrees in computer sciences, engineering and health sciences between 1975–76 and 1989–90. The Northeast saw the greatest growth in master's degrees in computer sciences with a 367 percent increase between 1975–76 and 1989–90, over 100 percentage points more than each of the other regions. In 1989–90 the Northeast awarded over one-third more master's in computer science than the next highest region. The West and South led the country in growth in master's degrees in engineering with 61 percent and 69 percent increases, respectively, between 1975–76 and 1989–90. There were increases in master's in health sciences in every region, ranging from 57 percent in the Midwest to 81 percent in the South.

Between 1985–86 and 1989–90, there was no clear trend in master's degrees as there was at the bachelor's level. Numbers of master's degrees conferred in computer sciences, engineering, and health sciences increased in every region. In the other science fields, master's degrees had small increases or declined in every region. The gap between the West and the other regions in the number of master's degrees awarded narrowed between 1985–86 and 1989–90 in health sciences, but widened in life sciences.

States

Tables 5–13 through 5–18 present state-by-state data on master's degrees by field. Master's degrees conferred in computer sciences, engineering, and

Text table 5.—Master's degrees by science field and region: 1975-76 to 1989-90

Field and region	1975-76	1980-81	1985-86	1989-90	Percent change 1975-76 to 1989-90	Percent change 1975-76 to 1980-81	Percent change 1980-81 to 1985-86	Percent change 1985-86 to 1989-90
Agricultural sciences								
United States	3,340	4,003	3,801	3,373	1.0	19.9	-5.0	-11.3
Northeast	431	427	423	435	0.9	-0.9	-0.9	2.8
Midwest	950	1,119	1,100	1,017	7.1	17.8	-1.7	-7.5
South	1,066	1,381	1,281	1,042	-2.3	29.5	-7.2	-18.7
West	893	1,076	997	879	-1.6	20.5	-7.3	-11.8
U.S.S.S.	—	—	—	—	(1)	(1)	(1)	(1)
Engineering								
United States	16,342	16,709	21,661	24,848	52.0	2.2	29.6	14.7
Northeast	4,653	4,604	5,606	6,449	38.6	-1.1	21.8	15.0
Midwest	3,787	3,871	5,068	5,662	49.5	2.2	30.9	11.7
South	3,887	4,160	5,730	6,580	69.3	7.0	37.7	14.8
West	3,684	3,723	4,778	5,946	61.4	1.1	28.3	24.4
U.S.S.S.	331	351	479	211	-36.3	6.0	36.5	-55.9
Life sciences								
United States	6,582	5,978	5,013	4,861	-26.1	-9.2	-16.1	-3.0
Northeast	1,659	1,619	1,409	1,407	-15.2	-2.4	-13.0	-0.1
Midwest	1,865	1,494	1,261	1,233	-33.9	-19.9	-15.6	-2.2
South	1,913	1,723	1,313	1,322	-30.9	-9.9	-23.8	0.7
West	1,145	1,141	1,018	899	-21.5	-0.3	-10.8	-11.7
U.S.S.S.	0	1	12	—	(1)	(1)	1100.0	(1)
Total sciences								
United States	46,218	52,196	63,071	68,526	48.3	12.9	20.8	8.6
Northeast	12,266	13,459	16,991	18,640	52.0	9.7	26.2	9.7
Midwest	11,831	13,132	15,322	16,157	36.6	11.0	16.7	5.4
South	12,083	14,119	17,563	18,769	55.3	16.9	24.4	6.9
West	9,587	11,002	12,591	14,650	52.8	14.8	14.4	16.4
U.S.S.S.	451	484	604	310	-31.3	7.3	24.8	-48.7
Computer sciences								
United States	2,603	4,218	8,070	9,643	270.5	62.0	91.3	19.5
Northeast	752	1,279	3,074	3,513	367.2	70.1	140.3	14.3
Midwest	598	1,001	1,470	1,783	198.2	67.4	46.9	21.3
South	711	1,070	2,242	2,570	261.5	50.5	109.5	14.6
West	484	793	1,232	1,678	246.7	63.8	55.4	36.2
U.S.S.S.	58	75	52	99	70.7	29.3	-30.7	90.4
Health sciences								
United States	11,885	16,004	18,624	20,354	71.3	34.7	16.4	9.3
Northeast	3,159	4,245	5,041	5,571	76.4	34.4	18.8	10.5
Midwest	3,237	4,327	4,968	5,087	57.2	33.7	14.8	2.4
South	3,209	4,458	5,459	5,817	81.3	38.9	22.5	6.6
West	2,280	2,974	3,156	3,879	70.1	30.4	6.1	22.9
U.S.S.S.	—	—	0	—	(1)	(1)	(1)	(1)
Physical sciences								
United States	5,466	5,284	5,902	5,447	-0.3	-3.3	11.7	-7.7
Northeast	1,612	1,285	1,438	1,265	-21.5	-20.3	11.9	-12.0
Midwest	1,394	1,320	1,455	1,375	-1.4	-5.3	10.2	-5.5
South	1,297	1,327	1,538	1,438	10.9	2.3	15.9	-6.5
West	1,101	1,295	1,410	1,369	24.3	17.6	8.9	-2.9
U.S.S.S.	62	57	61	0	-100.0	-8.1	7.0	-100.0

[1] Insufficient data for calculating a percent change.

—Data not available.

Note: U.S.S.S.=U. S. Service Schools.

SOURCE: U.S. Department of Education, National Center for Education Statistics, HEGIS, "Degrees and Other Formal Awards Conferred" surveys; and IPEDS, "Completions" surveys.

health sciences all show similar patterns when studied on a state-by-state basis. All three fields increased in most states between 1975–76 and 1989–90. However, most of the growth in computer sciences and engineering occurred during the 1980s.

Computer sciences master's degrees increased in 44 states by 50 percent or more between 1975–76 and 1989–90. Thirty-eight of those states had very large increases of 100 percent or more. There were increases of 50 percent or more in master's degrees conferred in engineering in 31 states and the District of Columbia, and in health sciences in 35 states during this same period. Between 1975–76 and 1989–90 master's degrees in agricultural sciences and in physical sciences declined in 18 states, and in life sciences declined in 39 states and the District of Columbia. Between 1985–86 and 1989–90, there were generally smaller increases in master's degrees in computer sciences, engineering, and health sciences than in 1980–81 to 1985–86, and in some states there was a decline in such degrees. Over half the states in the country had decreases in the numbers of master's conferred during the late 1980s in agricultural sciences, life sciences, and physical sciences.

In 1989–90, New York conferred 17 percent of the master's degrees in computer sciences, the largest percentage, followed by California with 12 percent. California enjoyed a big advantage over New York in the percentage of the total master's degrees awarded in engineering (15 percent of the total, compared with 9 percent) and Massachusetts was third with 7 percent. California, New York, and Massachusetts were also the top three states in numbers of master's degrees conferred in health sciences in 1989–90, with 12, 10, and 7 percent of the total, respectively.

Doctor's Degrees

Regions

Doctor's degrees awarded rose between 1975–76 and 1989–90 in every science field, and in every region (text table 6). On the regional level the South experienced the greatest percentage increases in doctor's degrees conferred in most fields (except for health sciences and physical sciences) with increases ranging from 29 percent in life sciences to 234 percent in computer sciences. In actual numbers of doctor's degrees conferred in 1989–90, the South led in life sciences and health sciences, while the Midwest conferred the most doctor's degrees in agricultural sciences, computer sciences, and engineering, and the Northeast conferred the most doctor's degrees in physical sciences. The West had the fewest doctor's degrees awarded in health sciences in 1975–76, but moved to second place in 1989–90 due to an increase of over 350 percent over the 15–year period.

States

Tables 5–19 through 5–24 present state-by-state data on doctor's degrees, by field. Overall, when states did offer doctor's degrees in a science field, the numbers rose in most states between 1975–76 and 1989–90.

California awarded far more doctor's degrees in health sciences in 1989–90 than any other state, accounting for over 20 percent of the total degrees and awarding more than the entire Northeast region, and almost as many as the Midwest.

Summary

There was growth on the national level between 1975–76 and 1989–90 in bachelor's and graduate degrees in computer sciences, engineering, and health sciences, but there were declines in bachelor's degrees in agricultural sciences, life sciences, and physical sciences. The picture changed during the late 1980s with decreases in the number of bachelor's degrees in all the science fields nationally, as well as in most regions and states. The rise and fall in the numbers of bachelor's degrees in computer sciences and engineering is the major story of the science fields.

Some areas for further research might include looking at how some institutions may attract increasing numbers of science graduates in the future, and how local industries might work with institutions to promote particular fields of study.

Text table 6.—Doctor's degrees by science field and region: 1975-76 to 1989-90

Field and region	1975-76	1980-81	1985-86	1989-90	Percent change 1975-76 to 1989-90	Percent change 1975-76 to 1980-81	Percent change 1980-81 to 1985-86	Percent change 1985-86 to 1989-90
Agricultural sciences								
United States	928	1,067	1,158	1,272	37.1	15.0	8.5	9.8
Northeast	129	110	120	149	15.5	-14.7	9.1	24.2
Midwest	368	426	420	444	20.7	15.8	-1.4	5.7
South	241	268	387	411	70.5	11.2	44.4	6.2
West	190	263	231	268	41.1	38.4	-12.2	16.0
U.S.S.S.	—	—	—	—	(1)	(1)	(1)	(1)
Engineering								
United States	2,821	2,561	3,410	4,965	76.0	-9.2	33.2	45.6
Northeast	775	704	906	1,297	67.4	-9.2	28.7	43.2
Midwest	720	742	952	1,321	83.5	3.1	28.3	38.8
South	604	479	764	1,210	100.3	-20.7	59.5	58.4
West	696	626	779	1,137	63.4	-10.1	24.4	46.0
U.S.S.S	26	10	9	0	-100.0	-61.5	-10.0	-100.0
Life sciences								
United States	3,392	3,718	3,358	3,844	13.3	9.6	-9.7	14.5
Northeast	903	931	836	1,023	13.3	3.1	-10.2	22.4
Midwest	919	948	807	947	3.0	3.2	-14.9	17.3
South	842	994	873	1,088	29.2	18.1	-12.2	24.6
West	728	845	834	786	8.0	16.1	-1.3	-5.8
U.S.S.S	0	0	8	—	(1)	(1)	(1)	(1)
Total science								
United States	11,393	11,566	13,062	16,415	44.1	1.5	12.9	25.7
Northeast	3,005	2,974	3,285	4,092	36.2	-1.0	10.5	24.6
Midwest	3,179	3,212	3,443	4,223	32.8	1.0	7.2	22.7
South	2,643	2,634	3,274	4,311	63.1	-0.3	24.3	31.7
West	2,538	2,735	3,041	3,789	49.3	7.8	11.2	24.6
U.S.S.S	28	11	19	0	-100.0	-60.7	72.7	-100.0
Computer sciences								
United States	244	252	344	623	155.3	3.3	36.5	81.1
Northeast	60	69	90	153	155.0	15.0	30.4	70.0
Midwest	89	66	95	183	105.6	-25.8	43.9	92.6
South	47	45	76	157	234.0	-4.3	68.9	106.6
West	48	72	83	130	170.8	50.0	15.3	56.6
U.S.S.S	0	0	0	—	(1)	(1)	(1)	(1)
Health sciences								
United States	577	827	1,241	1,543	167.4	43.3	50.1	24.3
Northeast	113	231	319	315	178.8	104.4	38.1	-1.3
Midwest	194	251	303	339	74.7	29.4	20.7	11.9
South	179	246	417	471	163.1	37.4	69.5	12.9
West	91	99	202	418	359.3	8.8	104.0	106.9
U.S.S.S	—	—	0	—	(1)	(1)	(1)	(1)
Physical sciences								
United States	3,431	3,141	3,551	4,168	21.5	-8.5	13.1	17.4
Northeast	1,025	929	1,014	1,155	12.7	-9.4	9.1	13.9
Midwest	889	779	866	989	11.2	-12.4	11.2	14.2
South	730	602	757	974	33.4	-17.5	25.7	28.7
West	785	830	912	1,050	33.8	5.7	9.9	15.1
U.S.S.S	2	1	2	0	-100.0	-50.0	100.0	-100.0

[1] Insufficient data for calculating a percent change.

—Data not available.

Note: U.S.S.S.=U. S. Service Schools.

SOURCE: U.S. Department of Education, National Center for Education Statistics, HEGIS, "Degrees and Other Formal Awards Conferred" surveys; and IPEDS, "Completions" surveys.

Table 5-1.—Associate degrees conferred in agricultural sciences, by region and state: 1982-83 to 1989-90

Region and state	1975-76	1976-77	1977-78	1978-79	1979-80	1980-81	1981-82	1982-83	1983-84	1984-85	1985-86	1986-87	1987-88	1988-89	1989-90	Percent change 1982-83 to 1989-90	Percent change 1982-83 to 1985-86	Percent change 1985-86 to 1989-90
United States	—	—	—	—	—	—	—	7,645	6,870	6,554	5,741	5,460	5,029	4,725	4,832	-36.8	-24.9	-15.8
Northeast	—	—	—	—	—	—	—	1,742	1,493	1,325	1,226	1,153	1,048	980	1,149	-34.0	-29.6	-6.3
Connecticut	—	—	—	—	—	—	—	0	1	0	0	0	1	1	0	([1])	([1])	([1])
Maine	—	—	—	—	—	—	—	67	41	48	34	26	15	26	36	-46.3	-49.3	5.9
Massachusetts	—	—	—	—	—	—	—	225	198	227	187	165	163	140	187	-16.9	-16.9	0.0
New Hampshire	—	—	—	—	—	—	—	85	98	60	63	68	67	60	62	-27.1	-25.9	-1.6
New Jersey	—	—	—	—	—	—	—	85	72	55	59	53	41	43	51	-40.0	-30.6	-13.6
New York	—	—	—	—	—	—	—	988	862	746	670	649	577	542	626	-36.6	-32.2	-6.6
Pennsylvania	—	—	—	—	—	—	—	248	177	149	178	157	131	118	121	-51.2	-28.2	-32.0
Rhode Island	—	—	—	—	—	—	—	9	15	0	0	—	15	16	21	133.3	-100.0	([1])
Vermont	—	—	—	—	—	—	—	35	29	40	35	35	38	34	45	28.6	0.0	28.6
Midwest	—	—	—	—	—	—	—	2,782	2,605	2,497	2,087	2,178	2,054	1,747	1,715	-38.4	-25.0	-17.8
Illinois	—	—	—	—	—	—	—	566	496	448	488	372	375	321	258	-54.4	-13.8	-47.1
Indiana	—	—	—	—	—	—	—	102	63	109	87	104	92	56	58	-43.1	-14.7	-33.3
Iowa	—	—	—	—	—	—	—	457	455	451	374	319	326	268	339	-25.8	-18.2	-9.4
Kansas	—	—	—	—	—	—	—	153	158	142	100	116	126	138	130	-15.0	-34.6	30.0
Michigan	—	—	—	—	—	—	—	130	114	66	64	56	67	60	55	-57.7	-50.8	-14.1
Minnesota	—	—	—	—	—	—	—	343	298	285	195	294	261	232	220	-35.9	-43.1	12.8
Missouri	—	—	—	—	—	—	—	98	118	117	89	77	74	48	53	-45.9	-9.2	-40.4
Nebraska	—	—	—	—	—	—	—	218	246	274	185	224	147	77	86	-60.6	-15.1	-53.5
North Dakota	—	—	—	—	—	—	—	157	167	137	114	123	109	71	79	-49.7	-27.4	-30.7
Ohio	—	—	—	—	—	—	—	390	267	276	228	347	352	326	290	-25.6	-41.5	27.2
South Dakota	—	—	—	—	—	—	—	27	32	20	13	7	9	12	10	-63.0	-51.9	-23.1
Wisconsin	—	—	—	—	—	—	—	141	191	172	150	139	116	138	137	-2.8	6.4	-8.7
South	—	—	—	—	—	—	—	1,501	1,362	1,320	1,208	1,035	1,049	1,124	1,065	-29.0	-19.5	-11.8
Alabama	—	—	—	—	—	—	—	28	20	45	31	41	47	25	28	0.0	10.7	-9.7
Arkansas	—	—	—	—	—	—	—	2	4	1	5	1	0	0	0	-100.0	150.0	-100.0
Delaware	—	—	—	—	—	—	—	17	14	13	8	7	6	9	4	-76.5	-52.9	-50.0
District of Columbia	—	—	—	—	—	—	—	0	1	0	1	—	—	—	—	([1])	([1])	([1])
Florida	—	—	—	—	—	—	—	137	135	135	135	73	84	92	78	-43.1	-1.5	-42.2
Georgia	—	—	—	—	—	—	—	146	138	124	110	99	107	83	98	-32.9	-24.7	-10.9
Kentucky	—	—	—	—	—	—	—	58	35	31	28	22	12	16	17	-70.7	-51.7	-39.3
Louisiana	—	—	—	—	—	—	—	14	16	21	12	9	4	8	4	-71.4	-14.3	-66.7
Maryland	—	—	—	—	—	—	—	19	23	14	17	15	13	18	16	-15.8	-10.5	-5.9
Mississippi	—	—	—	—	—	—	—	47	55	44	62	35	34	41	52	10.6	31.9	-16.1
North Carolina	—	—	—	—	—	—	—	355	326	269	244	253	237	235	231	-34.9	-31.3	-5.3
Oklahoma	—	—	—	—	—	—	—	54	121	149	130	137	113	132	137	153.7	140.7	5.4
South Carolina	—	—	—	—	—	—	—	80	69	72	81	43	39	43	49	-38.8	1.3	-39.5
Tennessee	—	—	—	—	—	—	—	17	31	21	21	17	13	9	27	58.8	23.5	28.6
Texas	—	—	—	—	—	—	—	243	219	231	212	168	219	218	169	-30.5	-12.8	-20.3
Virginia	—	—	—	—	—	—	—	250	123	115	82	97	94	159	128	-48.8	-67.2	56.1
West Virginia	—	—	—	—	—	—	—	34	32	35	29	18	27	36	27	-20.6	-14.7	-6.9
West	—	—	—	—	—	—	—	1,620	1,410	1,412	1,220	1,094	878	874	903	-44.3	-24.7	-26.0
Alaska	—	—	—	—	—	—	—	5	18	9	13	10	1	2	2	-60.0	160.0	-84.6
Arizona	—	—	—	—	—	—	—	71	33	56	41	30	15	20	11	-84.5	-42.3	-73.2
California	—	—	—	—	—	—	—	772	701	677	499	466	356	381	367	-52.5	-35.4	-26.5
Colorado	—	—	—	—	—	—	—	170	163	200	138	101	88	97	95	-44.1	-18.8	-31.2
Hawaii	—	—	—	—	—	—	—	19	22	9	14	13	18	5	8	-57.9	-26.3	-42.9
Idaho	—	—	—	—	—	—	—	102	71	91	64	58	76	51	75	-26.5	-37.3	17.2

Table 5-1.—Continued

Region and state	1975-76	1976-77	1977-78	1978-79	1979-80	1980-81	1981-82	1982-83	1983-84	1984-85	1985-86	1986-87	1987-88	1988-89	1989-90	Percent change 1982-83 to 1989-90	Percent change 1982-83 to 1985-86	Percent change 1985-86 to 1989-90
West Continued																		
Montana	—	—	—	—	—	—	—	20	18	24	29	31	30	22	17	-15.0	45.0	-41.4
Nevada	—	—	—	—	—	—	—	8	12	4	1	0	1	1	1	-87.5	-87.5	0.0
New Mexico	—	—	—	—	—	—	—	30	0	2	4	15	17	9	7	-76.7	-86.7	75.0
Oregon	—	—	—	—	—	—	—	209	154	141	115	100	79	86	94	-55.0	-45.0	-18.3
Utah	—	—	—	—	—	—	—	3	1	2	1	9	4	4	4	33.3	-66.7	300.0
Washington	—	—	—	—	—	—	—	161	117	97	196	180	103	99	136	-15.5	21.7	-30.6
Wyoming	—	—	—	—	—	—	—	50	100	100	105	81	90	97	86	72.0	110.0	-18.1
U.S. Service Schools	—	—	—	—	—	—	—	—	—	—	—	—	—	—	—	(1)	(1)	(1)
Outlying Areas	—	—	—	—	—	—	—	34	54	30	45	27	47	44	47	38.2	32.4	4.4
American Samoa	—	—	—	—	—	—	—	—	—	—	—	—	—	—	—	(1)	(1)	(1)
Guam	—	—	—	—	—	—	—	0	0	0	0	—	—	—	—	(1)	(1)	(1)
Northern Marianas	—	—	—	—	—	—	—	—	—	—	—	—	—	—	—	(1)	(1)	(1)
Puerto Rico	—	—	—	—	—	—	—	33	51	26	40	27	47	44	47	42.4	21.2	17.5
Trust Territories	—	—	—	—	—	—	—	0	3	3	2	—	0	—	—	(1)	(1)	(1)
Virgin Islands	—	—	—	—	—	—	—	1	0	1	3	—	—	—	—	(1)	200.0	(1)

[1] Insufficient data for calculating a percent change.

—Data not available.

SOURCE: U.S. Department of Education, National Center for Education Statistics, HEGIS, "Degrees and Other Formal Awards Conferred" surveys; and IPEDS, "Completions" surveys.

Table 5-2.—Associate degrees conferred in computer sciences, by region and state: 1982-83 to 1989-90

Region and state	1975-76	1976-77	1977-78	1978-79	1979-80	1980-81	1981-82	1982-83	1983-84	1984-85	1985-86	1986-87	1987-88	1988-89	1989-90	Percent change		
																1982-83 to 1989-90	1982-83 to 1985-86	1985-86 to 1989-90
United States	—	—	—	—	—	—	—	9,670	12,625	12,677	10,704	9,101	8,628	7,900	7,604	-21.4	10.7	-29.0
Northeast	—	—	—	—	—	—	—	1,928	2,161	2,020	1,735	1,522	2,190	1,809	1,735	-10.0	-10.0	0.0
Connecticut	—	—	—	—	—	—	—	0	0	0	0	0	60	51	39	([1])	([1])	([1])
Maine	—	—	—	—	—	—	—	27	32	146	59	53	47	31	28	3.7	118.5	-52.5
Massachusetts	—	—	—	—	—	—	—	670	692	521	488	414	360	296	250	-62.7	-27.2	-48.8
New Hampshire	—	—	—	—	—	—	—	142	190	238	226	148	113	87	87	-38.7	59.2	-61.5
New Jersey	—	—	—	—	—	—	—	321	273	277	230	186	167	169	141	-56.1	-28.3	-38.7
New York	—	—	—	—	—	—	—	0	0	0	0	—	980	799	765	([1])	([1])	([1])
Pennsylvania	—	—	—	—	—	—	—	554	596	546	459	439	295	244	269	-51.4	-17.1	-41.4
Rhode Island	—	—	—	—	—	—	—	156	305	222	221	258	147	114	136	-12.8	41.7	-38.5
Vermont	—	—	—	—	—	—	—	58	73	70	52	24	21	18	20	-65.5	-10.3	-61.5
Midwest	—	—	—	—	—	—	—	3,260	4,054	4,067	2,896	2,591	2,110	2,196	1,881	-42.3	-11.2	-35.0
Illinois	—	—	—	—	—	—	—	186	213	250	175	315	157	157	23	-87.6	-5.9	-86.9
Indiana	—	—	—	—	—	—	—	204	216	178	102	84	50	187	218	6.9	-50.0	113.7
Iowa	—	—	—	—	—	—	—	152	182	253	154	118	94	70	184	21.1	1.3	19.5
Kansas	—	—	—	—	—	—	—	181	248	248	143	125	85	85	95	-47.5	-21.0	-33.6
Michigan	—	—	—	—	—	—	—	887	969	1,101	867	709	582	580	487	-45.1	-2.3	-43.8
Minnesota	—	—	—	—	—	—	—	55	72	57	32	38	47	26	37	-32.7	-41.8	15.6
Missouri	—	—	—	—	—	—	—	139	301	273	207	272	215	139	114	-18.0	48.9	-44.9
Nebraska	—	—	—	—	—	—	—	38	39	22	22	80	73	50	86	126.3	-42.1	290.9
North Dakota	—	—	—	—	—	—	—	98	83	16	2	3	1	6	5	-94.9	-98.0	150.0
Ohio	—	—	—	—	—	—	—	1,051	1,332	1,258	867	604	525	657	442	-57.9	-17.5	-49.0
South Dakota	—	—	—	—	—	—	—	145	153	118	109	55	45	43	21	-85.5	-24.8	-80.7
Wisconsin	—	—	—	—	—	—	—	124	246	293	216	188	236	196	169	36.3	74.2	-21.8
South	—	—	—	—	—	—	—	2,207	3,332	3,351	3,088	2,602	2,362	2,122	2,118	-4.0	39.9	-31.4
Alabama	—	—	—	—	—	—	—	238	312	356	344	320	267	306	290	21.8	44.5	-15.7
Arkansas	—	—	—	—	—	—	—	52	100	30	30	98	102	54	26	-50.0	-42.3	-13.3
Delaware	—	—	—	—	—	—	—	87	123	100	97	76	78	39	45	-48.3	11.5	-53.6
District of Columbia	—	—	—	—	—	—	—	61	49	69	17	7	15	23	55	-9.8	-72.1	223.5
Florida	—	—	—	—	—	—	—	181	160	173	172	171	159	113	107	-40.9	-5.0	-37.8
Georgia	—	—	—	—	—	—	—	177	494	294	304	289	257	220	151	-14.7	71.8	-50.3
Kentucky	—	—	—	—	—	—	—	433	599	493	395	363	253	303	208	-52.0	-8.8	-47.3
Louisiana	—	—	—	—	—	—	—	35	124	146	210	121	58	45	44	25.7	500.0	-79.0
Maryland	—	—	—	—	—	—	—	0	0	0	31	60	63	66	106	([1])	([1])	241.9
Mississippi	—	—	—	—	—	—	—	303	346	356	257	185	137	126	147	-51.5	-15.2	-42.8
North Carolina	—	—	—	—	—	—	—	0	44	44	69	53	40	41	62	([1])	([1])	-10.1
Oklahoma	—	—	—	—	—	—	—	0	23	48	47	15	9	43	30	([1])	([1])	-36.2
South Carolina	—	—	—	—	—	—	—	7	7	12	21	21	21	2	4	-42.9	200.0	-81.0
Tennessee	—	—	—	—	—	—	—	180	253	243	231	229	249	174	143	-20.6	28.3	-38.1
Texas	—	—	—	—	—	—	—	386	516	772	684	424	436	448	497	28.8	77.2	-27.3
Virginia	—	—	—	—	—	—	—	9	58	52	53	71	112	80	169	1777.8	488.9	218.9
West Virginia	—	—	—	—	—	—	—	58	124	163	126	99	106	39	34	-41.4	117.2	-73.0
West	—	—	—	—	—	—	—	2,079	2,882	2,983	2,707	2,204	1,872	1,660	1,483	-28.7	30.2	-45.2
Alaska	—	—	—	—	—	—	—	14	13	9	1	—	—	—	0	-100.0	-92.9	-100.0
Arizona	—	—	—	—	—	—	—	213	425	533	417	264	207	232	143	-32.9	95.8	-65.7
California	—	—	—	—	—	—	—	1,292	1,655	1,576	1,303	1,166	913	831	775	-40.0	0.9	-40.5
Colorado	—	—	—	—	—	—	—	153	232	241	248	181	148	127	112	-26.8	62.1	-54.8
Hawaii	—	—	—	—	—	—	—	5	5	13	26	44	47	37	30	500.0	420.0	15.4
Idaho	—	—	—	—	—	—	—	28	47	55	58	54	37	40	37	32.1	107.1	-36.2

Table 5-2.—Continued

Region and state	1975-76	1976-77	1977-78	1978-79	1979-80	1980-81	1981-82	1982-83	1983-84	1984-85	1985-86	1986-87	1987-88	1988-89	1989-90	Percent change: 1982-83 to 1989-90	Percent change: 1982-83 to 1985-86	Percent change: 1985-86 to 1989-90
West Continued																		
Montana	—	—	—	—	—	—	—	10	13	14	9	9	18	15	21	110.0	-10.0	133.3
Nevada	—	—	—	—	—	—	—	6	42	0	51	40	31	37	19	216.7	750.0	-62.7
New Mexico	—	—	—	—	—	—	—	10	30	34	33	66	79	52	78	680.0	230.0	136.4
Oregon	—	—	—	—	—	—	—	44	56	53	25	15	36	53	41	-6.8	-43.2	64.0
Utah	—	—	—	—	—	—	—	51	106	138	151	76	102	20	23	-54.9	196.1	-84.8
Washington	—	—	—	—	—	—	—	223	194	260	361	268	236	198	180	-19.3	61.9	-50.1
Wyoming	—	—	—	—	—	—	—	30	64	57	24	21	18	18	24	-20.0	-20.0	0.0
U.S. Service Schools	—	—	—	—	—	—	—	196	196	256	278	182	94	113	387	97.4	41.8	39.2
Outlying Areas	—	—	—	—	—	—	—	248	288	213	258	185	246	278	236	-4.8	4.0	-8.5
American Samoa	—	—	—	—	—	—	—	—	—	—	—	—	—	—	—	([1])	([1])	([1])
Guam	—	—	—	—	—	—	—	1	9	4	8	4	7	11	8	700.0	700.0	0.0
Northern Marianas	—	—	—	—	—	—	—	—	—	—	—	—	—	—	—	([1])	([1])	([1])
Puerto Rico	—	—	—	—	—	—	—	247	279	203	250	181	239	267	228	-7.7	1.2	-8.8
Trust Territories	—	—	—	—	—	—	—	—	—	—	—	—	—	—	—	([1])	([1])	([1])
Virgin Islands	—	—	—	—	—	—	—	—	—	6	—	—	—	—	—	([1])	([1])	([1])

[1] Insufficient data for calculating a percent change.

—Data not available.

SOURCE: U.S. Department of Education, National Center for Education Statistics, HEGIS, "Degrees and Other Formal Awards Conferred" surveys; and IPEDS, "Completions" surveys.

Table 5-3.—Associate degrees conferred in engineering, by region and state: 1982-83 to 1989-90

Region and state	1975-76	1976-77	1977-78	1978-79	1979-80	1980-81	1981-82	1982-83	1983-84	1984-85	1985-86	1986-87	1987-88	1988-89	1989-90	Percent change 1982-83 to 1989-90	1982-83 to 1985-86	1985-86 to 1989-90
United States	—	—	—	—	—	—	—	60,629	61,699	63,832	63,339	62,510	62,227	56,368	54,131	-10.7	4.5	-14.5
Northeast	—	—	—	—	—	—	—	14,566	14,681	13,898	13,250	12,208	12,557	10,917	11,367	-22.0	-9.0	-14.2
Connecticut	—	—	—	—	—	—	—	828	935	897	791	689	614	616	637	-23.1	-4.5	-19.5
Maine	—	—	—	—	—	—	—	274	245	325	278	307	335	207	184	-32.8	1.5	-33.8
Massachusetts	—	—	—	—	—	—	—	2,041	2,077	1,803	1,655	1,383	1,505	1,399	1,370	-32.9	-18.9	-17.2
New Hampshire	—	—	—	—	—	—	—	423	525	591	547	447	425	399	357	-15.6	29.3	-34.7
New Jersey	—	—	—	—	—	—	—	685	664	699	659	651	624	560	526	-23.2	-3.8	-20.2
New York	—	—	—	—	—	—	—	5,444	5,239	5,009	4,778	4,464	4,416	4,150	4,252	-21.9	-12.2	-11.0
Pennsylvania	—	—	—	—	—	—	—	4,161	4,328	3,908	3,828	3,379	3,970	2,926	3,269	-21.4	-8.0	-14.6
Rhode Island	—	—	—	—	—	—	—	514	484	475	529	704	488	507	571	11.1	2.9	7.9
Vermont	—	—	—	—	—	—	—	196	184	191	185	184	180	153	201	2.6	-5.6	8.6
Midwest	—	—	—	—	—	—	—	16,588	17,665	18,118	17,334	17,941	18,200	16,600	14,834	-10.6	4.5	-14.4
Illinois	—	—	—	—	—	—	—	2,452	2,588	2,500	2,626	2,907	3,425	2,692	1,824	-25.6	7.1	-30.5
Indiana	—	—	—	—	—	—	—	2,394	2,894	2,947	2,627	2,658	2,896	2,396	2,223	-7.1	9.7	-15.4
Iowa	—	—	—	—	—	—	—	847	963	1,196	1,184	1,411	1,301	1,547	1,410	66.5	39.8	19.1
Kansas	—	—	—	—	—	—	—	287	325	303	313	332	275	303	312	8.7	9.1	-0.3
Michigan	—	—	—	—	—	—	—	3,102	3,175	3,100	2,714	2,725	2,521	2,213	2,450	-21.0	-12.5	-9.7
Minnesota	—	—	—	—	—	—	—	442	441	499	432	603	637	528	295	-33.3	-2.3	-31.7
Missouri	—	—	—	—	—	—	—	630	679	859	801	738	894	887	702	11.4	27.1	-12.4
Nebraska	—	—	—	—	—	—	—	931	858	881	792	795	839	761	734	-21.2	-14.9	-7.3
North Dakota	—	—	—	—	—	—	—	404	360	363	547	561	496	450	446	10.4	35.4	-18.5
Ohio	—	—	—	—	—	—	—	3,487	3,549	3,822	3,639	3,706	3,532	3,518	3,091	-11.4	4.4	-15.1
South Dakota	—	—	—	—	—	—	—	108	123	33	37	23	17	13	14	-87.0	-65.7	-62.2
Wisconsin	—	—	—	—	—	—	—	1,504	1,710	1,615	1,622	1,482	1,367	1,292	1,333	-11.4	7.8	-17.8
South	—	—	—	—	—	—	—	16,116	15,408	16,720	16,743	16,284	16,134	15,907	14,445	-10.4	3.9	-13.7
Alabama	—	—	—	—	—	—	—	167	206	844	802	1,073	930	696	770	361.1	380.2	-4.0
Arkansas	—	—	—	—	—	—	—	264	344	330	312	494	592	566	492	86.4	18.2	57.7
Delaware	—	—	—	—	—	—	—	215	236	216	207	212	167	162	205	-4.7	-3.7	-1.0
District of Columbia	—	—	—	—	—	—	—	88	72	75	70	69	84	55	59	-33.0	-20.5	-15.7
Florida	—	—	—	—	—	—	—	3,192	2,591	2,886	3,237	2,723	2,436	2,447	1,922	-39.8	1.4	-40.6
Georgia	—	—	—	—	—	—	—	444	471	492	524	629	551	692	541	21.8	18.0	3.2
Kentucky	—	—	—	—	—	—	—	608	601	1,011	838	612	628	531	589	-3.1	37.8	-29.7
Louisiana	—	—	—	—	—	—	—	420	471	456	414	458	390	429	283	-32.6	-1.4	-31.6
Maryland	—	—	—	—	—	—	—	383	390	395	366	358	366	330	353	-7.8	-4.4	-3.6
Mississippi	—	—	—	—	—	—	—	671	543	518	538	546	554	541	467	-30.4	-19.8	-13.2
North Carolina	—	—	—	—	—	—	—	1,429	1,522	1,418	1,461	1,403	1,361	1,205	1,202	-15.9	2.2	-17.7
Oklahoma	—	—	—	—	—	—	—	596	669	715	1,473	1,180	986	1,439	1,241	108.2	147.1	-15.8
South Carolina	—	—	—	—	—	—	—	806	944	859	812	734	800	802	804	-0.2	0.7	-1.0
Tennessee	—	—	—	—	—	—	—	1,308	1,142	1,096	1,007	827	907	811	784	-40.1	-23.0	-22.1
Texas	—	—	—	—	—	—	—	3,379	3,774	3,938	3,420	3,539	3,927	3,634	3,286	-2.8	1.2	-3.9
Virginia	—	—	—	—	—	—	—	1,692	1,042	985	812	1,007	1,081	929	891	-47.3	-52.0	9.7
West Virginia	—	—	—	—	—	—	—	454	390	486	450	420	374	638	556	22.5	-0.9	23.6
West	—	—	—	—	—	—	—	11,163	11,749	12,291	12,257	12,205	12,230	10,628	10,261	-8.1	9.8	-16.3
Alaska	—	—	—	—	—	—	—	149	119	104	98	112	93	78	58	-61.1	-34.2	-40.8
Arizona	—	—	—	—	—	—	—	1,056	804	855	1,333	603	1,202	1,335	1,040	-1.5	26.2	-22.0
California	—	—	—	—	—	—	—	4,732	5,343	5,870	5,203	5,894	5,626	4,467	4,162	-12.0	10.0	-20.0
Colorado	—	—	—	—	—	—	—	1,262	1,146	1,169	1,154	1,567	1,529	1,225	1,414	12.0	-8.6	22.5
Hawaii	—	—	—	—	—	—	—	481	462	451	435	387	333	287	275	-42.8	-9.6	-36.8
Idaho	—	—	—	—	—	—	—	268	312	346	299	282	327	378	409	52.6	11.6	36.8

Table 5-3.—Continued

Region and state	1975-76	1976-77	1977-78	1978-79	1979-80	1980-81	1981-82	1982-83	1983-84	1984-85	1985-86	1986-87	1987-88	1988-89	1989-90	Percent change 1982-83 to 1989-90	Percent change 1982-83 to 1985-86	Percent change 1985-86 to 1989-90
West Continued																		
Montana	—	—	—	—	—	—	—	137	137	138	106	91	105	58	73	-46.7	-22.6	-31.1
Nevada	—	—	—	—	—	—	—	110	123	139	109	66	81	59	60	-45.5	-0.9	-45.0
New Mexico	—	—	—	—	—	—	—	384	478	313	281	278	189	168	231	-39.8	-26.8	-17.8
Oregon	—	—	—	—	—	—	—	1,048	1,228	1,347	1,229	1,088	968	992	909	-13.3	17.3	-26.0
Utah.............	—	—	—	—	—	—	—	468	466	512	477	553	655	524	528	12.8	1.9	10.7
Washington.......	—	—	—	—	—	—	—	925	988	930	1,388	1,134	839	887	951	2.8	50.1	-31.5
Wyoming	—	—	—	—	—	—	—	143	143	117	145	150	283	170	151	5.6	1.4	4.1
U.S. Service Schools............	—	—	—	—	—	—	—	2,196	2,196	2,805	3,755	3,872	3,106	2,316	3,224	46.8	71.0	-14.1
Outlying Areas......	—	—	—	—	—	—	—	475	503	401	475	373	314	382	306	-35.6	0.0	-35.6
American Samoa ..	—	—	—	—	—	—	—	16	11	11	9	—	—	—	—	([1])	-43.8	([1])
Guam	—	—	—	—	—	—	—	8	4	9	6	8	12	7	5	-37.5	-25.0	-16.7
Northern Marianas.	—	—	—	—	—	—	—	—	—	—	—	—	—	—	—	([1])	([1])	([1])
Puerto Rico	—	—	—	—	—	—	—	451	479	372	454	365	292	365	301	-33.3	0.7	-33.7
Trust Territories ...	—	—	—	—	—	—	—	0	9	9	6	—	10	10	—	([1])	([1])	([1])
Virgin Islands	—	—	—	—	—	—	—	—	—	—	—	—	—	—	—	([1])	([1])	([1])

[1] Insufficient data for calculating a percent change.

—Data not available.

SOURCE: U.S. Department of Education, National Center for Education Statistics, HEGIS, "Degrees and Other Formal Awards Conferred" surveys; and IPEDS, "Completions" surveys.

Table 5-4.—Associate degrees conferred in health sciences, by region and state: 1982-83 to 1989-90

Region and state	1975-76	1976-77	1977-78	1978-79	1979-80	1980-81	1981-82	1982-83	1983-84	1984-85	1985-86	1986-87	1987-88	1988-89	1989-90	Percent change 1982-83 to 1989-90	Percent change 1982-83 to 1985-86	Percent change 1985-86 to 1989-90
United States	—	—	—	—	—	—	—	65,749	68,110	68,453	66,559	62,546	59,711	59,566	64,128	-2.5	1.2	-3.7
Northeast	—	—	—	—	—	—	—	14,495	15,023	15,294	14,824	14,278	13,270	13,144	14,141	-2.4	2.3	-4.6
Connecticut	—	—	—	—	—	—	—	734	659	733	595	596	592	544	587	-20.0	-18.9	-1.3
Maine	—	—	—	—	—	—	—	416	446	397	313	364	354	365	374	-10.1	-24.8	19.5
Massachusetts	—	—	—	—	—	—	—	2,272	2,308	2,350	2,239	2,104	1,876	1,880	1,892	-16.7	-1.5	-15.5
New Hampshire	—	—	—	—	—	—	—	260	308	368	319	348	326	353	390	50.0	22.7	22.3
New Jersey	—	—	—	—	—	—	—	1,783	1,998	1,896	1,940	1,659	1,547	1,439	1,619	-9.2	8.8	-16.5
New York	—	—	—	—	—	—	—	5,949	6,002	6,065	5,872	5,765	5,555	5,633	5,893	-0.9	-1.3	0.4
Pennsylvania	—	—	—	—	—	—	—	2,589	2,827	2,954	3,006	2,943	2,612	2,581	2,872	10.9	16.1	-4.5
Rhode Island	—	—	—	—	—	—	—	270	261	305	338	341	280	233	353	30.7	25.2	4.4
Vermont	—	—	—	—	—	—	—	222	214	226	202	158	128	116	161	-27.5	-9.0	-20.3
Midwest	—	—	—	—	—	—	—	17,804	19,095	19,339	18,879	17,798	16,595	16,456	17,381	-2.4	6.0	-7.9
Illinois	—	—	—	—	—	—	—	3,267	3,436	3,553	3,489	3,023	2,827	2,676	2,700	-17.4	6.8	-22.6
Indiana	—	—	—	—	—	—	—	1,760	1,720	1,665	1,737	1,504	1,341	1,461	1,615	-8.2	-1.3	-7.0
Iowa	—	—	—	—	—	—	—	1,006	1,089	1,019	1,072	998	977	1,159	1,192	18.5	6.6	11.2
Kansas	—	—	—	—	—	—	—	738	835	870	806	780	745	839	998	35.2	9.2	23.8
Michigan	—	—	—	—	—	—	—	3,298	3,621	3,900	3,607	3,787	3,285	2,847	2,763	-16.2	9.4	-23.4
Minnesota	—	—	—	—	—	—	—	1,315	1,376	1,399	1,178	1,288	1,228	1,099	1,382	5.1	-10.4	17.3
Missouri	—	—	—	—	—	—	—	886	1,044	1,077	1,016	951	877	888	829	-6.4	14.7	-18.4
Nebraska	—	—	—	—	—	—	—	411	460	365	337	206	140	218	257	-37.5	-18.0	-23.7
North Dakota	—	—	—	—	—	—	—	165	154	189	154	156	168	128	161	-2.4	-6.7	4.5
Ohio	—	—	—	—	—	—	—	3,507	3,675	3,731	3,909	3,666	3,504	3,693	3,858	10.0	11.5	-1.3
South Dakota	—	—	—	—	—	—	—	314	329	304	308	267	298	274	297	-5.4	-1.9	-3.6
Wisconsin	—	—	—	—	—	—	—	1,137	1,356	1,267	1,266	1,172	1,205	1,174	1,329	16.9	11.3	5.0
South	—	—	—	—	—	—	—	21,869	21,841	22,203	21,502	19,581	19,091	19,849	21,892	0.1	-1.7	1.8
Alabama	—	—	—	—	—	—	—	1,221	1,403	1,359	1,360	1,170	1,090	1,125	1,069	-12.4	11.4	-21.4
Arkansas	—	—	—	—	—	—	—	598	547	617	573	534	484	532	620	3.7	-4.2	8.2
Delaware	—	—	—	—	—	—	—	247	215	238	214	234	260	287	323	30.8	-13.4	50.9
District of Columbia	—	—	—	—	—	—	—	123	156	180	186	149	131	155	112	-8.9	51.2	-39.8
Florida	—	—	—	—	—	—	—	3,861	3,463	3,455	3,542	3,102	2,919	2,990	3,277	-15.1	-8.3	-7.5
Georgia	—	—	—	—	—	—	—	1,146	1,263	1,227	1,233	1,151	1,093	1,266	1,372	19.7	7.6	11.3
Kentucky	—	—	—	—	—	—	—	1,482	1,585	1,577	1,411	1,208	1,147	1,198	1,259	-15.0	-4.8	-10.8
Louisiana	—	—	—	—	—	—	—	492	440	496	526	494	514	487	553	12.4	6.9	5.1
Maryland	—	—	—	—	—	—	—	1,348	1,441	1,427	1,325	1,222	1,108	960	1,044	-22.6	-1.7	-21.2
Mississippi	—	—	—	—	—	—	—	979	989	1,053	929	867	849	902	912	-6.8	-5.1	-1.8
North Carolina	—	—	—	—	—	—	—	1,663	2,019	1,967	1,887	1,791	1,628	1,687	1,984	19.3	13.5	5.1
Oklahoma	—	—	—	—	—	—	—	671	718	674	615	617	711	815	903	34.6	-8.3	46.8
South Carolina	—	—	—	—	—	—	—	793	889	901	895	824	779	807	932	17.5	12.9	4.1
Tennessee	—	—	—	—	—	—	—	1,464	1,417	1,524	1,398	1,364	1,323	1,279	1,383	-5.5	-4.5	-1.1
Texas	—	—	—	—	—	—	—	2,844	3,333	3,416	3,343	3,059	3,219	3,542	4,069	43.1	17.5	21.7
Virginia	—	—	—	—	—	—	—	2,203	1,165	1,221	1,214	1,080	1,230	1,224	1,382	-37.3	-44.9	13.8
West Virginia	—	—	—	—	—	—	—	734	798	871	851	715	606	593	698	-4.9	15.9	-18.0
West	—	—	—	—	—	—	—	11,210	11,780	11,351	10,905	10,452	10,175	9,722	10,227	-8.8	-2.7	-6.2
Alaska	—	—	—	—	—	—	—	81	54	90	77	89	84	83	91	12.3	-4.9	18.2
Arizona	—	—	—	—	—	—	—	743	689	740	750	676	725	721	734	-1.2	0.9	-2.1
California	—	—	—	—	—	—	—	6,039	6,343	6,287	5,409	5,172	4,789	4,460	4,395	-27.2	-10.4	-18.7
Colorado	—	—	—	—	—	—	—	512	583	567	645	705	665	628	719	40.4	26.0	11.5
Hawaii	—	—	—	—	—	—	—	141	160	151	168	166	163	190	192	36.2	19.1	14.3
Idaho	—	—	—	—	—	—	—	349	350	305	300	264	302	298	345	-1.1	-14.0	15.0

Table 5-4.—Continued

Region and state	1975-76	1976-77	1977-78	1978-79	1979-80	1980-81	1981-82	1982-83	1983-84	1984-85	1985-86	1986-87	1987-88	1988-89	1989-90	Percent change: 1982-83 to 1989-90	Percent change: 1982-83 to 1985-86	Percent change: 1985-86 to 1989-90
West Continued																		
Montana	—	—	—	—	—	—	—	71	107	96	94	93	98	111	134	88.7	32.4	42.6
Nevada	—	—	—	—	—	—	—	168	216	166	183	167	196	195	168	0.0	8.9	-8.2
New Mexico	—	—	—	—	—	—	—	400	415	447	467	348	399	346	527	31.8	16.8	12.8
Oregon	—	—	—	—	—	—	—	850	827	779	851	805	845	762	809	-4.8	0.1	-4.9
Utah	—	—	—	—	—	—	—	516	508	372	311	337	371	461	461	-10.7	-39.7	48.2
Washington	—	—	—	—	—	—	—	1,226	1,359	1,184	1,472	1,482	1,339	1,262	1,455	18.7	20.1	-1.2
Wyoming	—	—	—	—	—	—	—	114	169	167	178	148	199	205	197	72.8	56.1	10.7
U.S. Service Schools	—	—	—	—	—	—	—	371	371	266	449	437	580	395	487	31.3	21.0	8.5
Outlying Areas	—	—	—	—	—	—	—	882	1,032	705	796	787	732	719	801	-9.2	-9.8	0.6
American Samoa	—	—	—	—	—	—	—	4	21	21	4	—	—	—	—	(1)	0.0	(1)
Guam	—	—	—	—	—	—	—	13	32	14	21	18	18	21	19	46.2	61.5	-9.5
Northern Marianas	—	—	—	—	—	—	—	—	—	—	—	—	—	—	1	(1)	(1)	(1)
Puerto Rico	—	—	—	—	—	—	—	855	971	658	771	765	706	689	769	-10.1	-9.8	-0.3
Trust Territories	—	—	—	—	—	—	—	—	—	—	—	—	—	—	10	(1)	(1)	(1)
Virgin Islands	—	—	—	—	—	—	—	10	8	12	—	4	8	9	2	-80.0	(1)	(1)

[1] Insufficient data for calculating a percent change.

—Data not available.

SOURCE: U.S. Department of Education, National Center for Education Statistics, HEGIS, "Degrees and Other Formal Awards Conferred" surveys; and IPEDS, "Completions" surveys.

Table 5-5.—Associate degrees conferred in life sciences, by region and state: 1982-83 to 1989-90

Region and state	1975-76	1976-77	1977-78	1978-79	1979-80	1980-81	1981-82	1982-83	1983-84	1984-85	1985-86	1986-87	1987-88	1988-89	1989-90	Percent change 1982-83 to 1989-90	Percent change 1982-83 to 1985-86	Percent change 1985-86 to 1989-90
United States	—	—	—	—	—	—	—	981	1,209	1,121	998	893	854	982	1,034	5.4	1.7	3.6
Northeast	—	—	—	—	—	—	—	101	97	108	104	99	104	113	86	-14.9	3.0	-17.3
Connecticut	—	—	—	—	—	—	—	0	0	0	0	—	—	—	—	(1)	(1)	(1)
Maine	—	—	—	—	—	—	—	0	0	0	0	—	—	—	—	(1)	(1)	(1)
Massachusetts	—	—	—	—	—	—	—	1	0	0	0	—	—	—	—	(1)	(1)	(1)
New Hampshire	—	—	—	—	—	—	—	3	7	1	1	—	—	—	1	-66.7	-66.7	0.0
New Jersey	—	—	—	—	—	—	—	21	16	29	19	31	31	29	23	9.5	-9.5	21.1
New York	—	—	—	—	—	—	—	30	20	24	32	25	22	27	27	-10.0	6.7	-15.6
Pennsylvania	—	—	—	—	—	—	—	46	54	52	45	37	48	56	33	-28.3	-2.2	-26.7
Rhode Island	—	—	—	—	—	—	—	0	0	0	7	6	3	1	2	(1)	(1)	-71.4
Vermont	—	—	—	—	—	—	—	0	0	2	0	0	—	—	—	(1)	(1)	(1)
Midwest	—	—	—	—	—	—	—	233	231	212	180	130	133	87	100	-57.1	-22.7	-44.4
Illinois	—	—	—	—	—	—	—	122	79	89	7	5	5	5	—	(1)	-94.3	(1)
Indiana	—	—	—	—	—	—	—	3	1	0	2	—	—	2	1	-66.7	-33.3	-50.0
Iowa	—	—	—	—	—	—	—	0	1	5	0	—	—	—	—	(1)	(1)	(1)
Kansas	—	—	—	—	—	—	—	23	31	16	19	25	29	29	56	143.5	-17.4	194.7
Michigan	—	—	—	—	—	—	—	54	66	61	120	81	82	33	29	-46.3	122.2	-75.8
Minnesota	—	—	—	—	—	—	—	5	3	2	0	0	—	—	—	(1)	-100.0	(1)
Missouri	—	—	—	—	—	—	—	15	3	6	1	5	4	4	7	-53.3	-93.3	600.0
Nebraska	—	—	—	—	—	—	—	6	10	2	12	2	1	3	1	-83.3	100.0	-91.7
North Dakota	—	—	—	—	—	—	—	2	2	2	0	1	1	—	—	(1)	-100.0	(1)
Ohio	—	—	—	—	—	—	—	2	33	28	19	11	11	11	6	200.0	850.0	-68.4
South Dakota	—	—	—	—	—	—	—	1	2	1	0	—	—	—	—	(1)	-100.0	(1)
Wisconsin	—	—	—	—	—	—	—	0	0	0	0	0	—	—	—	(1)	(1)	(1)
South	—	—	—	—	—	—	—	143	254	255	235	221	206	178	173	21.0	64.3	-26.4
Alabama	—	—	—	—	—	—	—	8	5	3	17	4	3	9	5	-37.5	112.5	-70.6
Arkansas	—	—	—	—	—	—	—	0	0	0	0	—	—	—	—	(1)	(1)	(1)
Delaware	—	—	—	—	—	—	—	4	4	0	0	1	0	0	0	(1)	-100.0	(1)
District of Columbia	—	—	—	—	—	—	—	4	0	14	0	2	7	3	—	(1)	-100.0	(1)
Florida	—	—	—	—	—	—	—	0	0	0	1	—	3	0	0	(1)	(1)	-100.0
Georgia	—	—	—	—	—	—	—	0	7	5	1	2	3	1	—	(1)	(1)	(1)
Kentucky	—	—	—	—	—	—	—	3	3	8	6	6	2	2	2	-33.3	100.0	-66.7
Louisiana	—	—	—	—	—	—	—	5	4	2	8	2	3	7	4	-20.0	60.0	-50.0
Maryland	—	—	—	—	—	—	—	0	0	0	0	—	—	—	—	(1)	(1)	(1)
Mississippi	—	—	—	—	—	—	—	16	12	21	15	26	17	28	18	12.5	-6.3	20.0
North Carolina	—	—	—	—	—	—	—	0	0	11	0	1	—	—	3	(1)	(1)	(1)
Oklahoma	—	—	—	—	—	—	—	3	117	125	122	114	101	61	63	2000.0	3966.7	-48.4
South Carolina	—	—	—	—	—	—	—	0	0	0	0	0	0	—	—	(1)	(1)	(1)
Tennessee	—	—	—	—	—	—	—	20	13	4	6	6	5	7	5	-75.0	-70.0	-16.7
Texas	—	—	—	—	—	—	—	65	77	50	50	54	54	56	68	4.6	-23.1	36.0
Virginia	—	—	—	—	—	—	—	15	12	12	8	3	8	4	5	-66.7	-46.7	-37.5
West Virginia	—	—	—	—	—	—	—	0	0	0	1	—	—	—	—	(1)	(1)	(1)
West	—	—	—	—	—	—	—	504	627	546	479	443	411	604	675	33.9	-5.0	40.9
Alaska	—	—	—	—	—	—	—	3	11	1	0	—	—	—	—	(1)	-100.0	(1)
Arizona	—	—	—	—	—	—	—	1	47	1	0	2	0	1	0	-100.0	-100.0	(1)
California	—	—	—	—	—	—	—	439	494	487	425	394	360	536	610	39.0	-3.2	43.5
Colorado	—	—	—	—	—	—	—	13	4	3	3	—	—	—	—	(1)	-76.9	(1)
Hawaii	—	—	—	—	—	—	—	0	0	0	0	—	—	—	—	(1)	(1)	(1)
Idaho	—	—	—	—	—	—	—	11	15	20	14	23	23	27	28	154.5	27.3	100.0

Table 5-5.—Continued

Region and state	1975-76	1976-77	1977-78	1978-79	1979-80	1980-81	1981-82	1982-83	1983-84	1984-85	1985-86	1986-87	1987-88	1988-89	1989-90	Percent change 1982-83 to 1989-90	Percent change 1982-83 to 1985-86	Percent change 1985-86 to 1989-90
West Continued																		
Montana	—	—	—	—	—	—	—	0	0	0	0	2	2	2	2	([1])	([1])	([1])
Nevada	—	—	—	—	—	—	—	0	0	0	0	—	—	—	—	([1])	([1])	([1])
New Mexico	—	—	—	—	—	—	—	2	1	1	0	—	—	1	1	-50.0	-100.0	([1])
Oregon	—	—	—	—	—	—	—	2	1	3	3	2	—	6	1	-50.0	50.0	-66.7
Utah	—	—	—	—	—	—	—	0	0	0	0	—	1	7	12	([1])	([1])	([1])
Washington	—	—	—	—	—	—	—	18	26	6	9	—	—	—	—	([1])	-50.0	([1])
Wyoming	—	—	—	—	—	—	—	15	28	24	25	20	25	24	21	40.0	66.7	-16.0
U.S. Service Schools	—	—	—	—	—	—	—	0	0	0	0	—	—	—	—	([1])	([1])	([1])
Outlying Areas	—	—	—	—	—	—	—	9	185	112	97	102	83	29	21	133.3	977.8	-78.4
American Samoa	—	—	—	—	—	—	—	—	—	—	—	—	—	—	—	([1])	([1])	([1])
Guam	—	—	—	—	—	—	—	0	0	0	0	—	—	—	—	([1])	([1])	([1])
Northern Marianas	—	—	—	—	—	—	—	—	—	—	—	—	—	—	—	([1])	([1])	([1])
Puerto Rico	—	—	—	—	—	—	—	9	185	112	86	102	83	29	21	133.3	855.6	-75.6
Trust Territories	—	—	—	—	—	—	—	—	—	—	—	—	—	—	—	([1])	([1])	([1])
Virgin Islands	—	—	—	—	—	—	—	0	0	0	11	—	—	—	—	([1])	([1])	([1])

[1] Insufficient data for calculating a percent change.

—Data not available.

SOURCE: U.S. Department of Education, National Center for Education Statistics, HEGIS, "Degrees and Other Formal Awards Conferred" surveys; and IPEDS, "Completions" surveys.

Table 5-6.—Associate degrees conferred in physical sciences, by region and state: 1982-83 to 1989-90

Region and state	1975-76	1976-77	1977-78	1978-79	1979-80	1980-81	1981-82	1982-83	1983-84	1984-85	1985-86	1986-87	1987-88	1988-89	1989-90	Percent change 1982-83 to 1989-90	Percent change 1982-83 to 1985-86	Percent change 1985-86 to 1989-90
United States	—	—	—	—	—	—	—	3,103	2,877	2,193	2,107	2,061	1,890	1,961	2,135	-31.2	-32.1	1.3
Northeast	—	—	—	—	—	—	—	721	721	598	543	512	375	374	362	-49.8	-24.7	-33.3
Connecticut	—	—	—	—	—	—	—	20	23	22	36	50	48	69	50	150.0	80.0	38.9
Maine	—	—	—	—	—	—	—	9	10	8	7	5	6	—	4	-55.6	-22.2	-42.9
Massachusetts	—	—	—	—	—	—	—	13	16	21	12	6	6	9	11	-15.4	-7.7	-8.3
New Hampshire	—	—	—	—	—	—	—	2	1	2	0	—	—	—	—	(1)	-100.0	(1)
New Jersey	—	—	—	—	—	—	—	113	88	68	84	63	62	57	52	-54.0	-25.7	-38.1
New York	—	—	—	—	—	—	—	447	446	373	312	306	174	185	171	-61.7	-30.2	-45.2
Pennsylvania	—	—	—	—	—	—	—	100	118	92	73	78	68	44	69	-31.0	-27.0	-5.5
Rhode Island	—	—	—	—	—	—	—	13	19	12	19	4	11	10	5	-61.5	46.2	-73.7
Vermont	—	—	—	—	—	—	—	4	0	0	0	0	—	—	—	(1)	-100.0	(1)
Midwest	—	—	—	—	—	—	—	746	724	479	440	442	427	448	456	-38.9	-41.0	3.6
Illinois	—	—	—	—	—	—	—	45	21	30	23	19	20	11	13	-71.1	-48.9	-43.5
Indiana	—	—	—	—	—	—	—	77	60	57	51	46	31	26	30	-61.0	-33.8	-41.2
Iowa	—	—	—	—	—	—	—	0	0	2	1	—	1	5	6	(1)	(1)	500.0
Kansas	—	—	—	—	—	—	—	27	18	11	8	19	20	17	17	-37.0	-70.4	112.5
Michigan	—	—	—	—	—	—	—	459	495	247	255	255	255	290	282	-38.6	-44.4	10.6
Minnesota	—	—	—	—	—	—	—	5	5	6	2	2	—	—	—	(1)	-60.0	(1)
Missouri	—	—	—	—	—	—	—	17	3	17	11	10	6	11	2	-88.2	-35.3	-81.8
Nebraska	—	—	—	—	—	—	—	10	5	2	0	—	—	6	5	-50.0	-100.0	(1)
North Dakota	—	—	—	—	—	—	—	0	0	5	3	4	2	—	7	(1)	(1)	133.3
Ohio	—	—	—	—	—	—	—	74	83	67	53	62	70	61	64	-13.5	-28.4	20.8
South Dakota	—	—	—	—	—	—	—	18	13	0	0	—	—	—	—	(1)	-100.0	(1)
Wisconsin	—	—	—	—	—	—	—	14	21	35	33	25	22	21	30	114.3	135.7	-9.1
South	—	—	—	—	—	—	—	859	640	536	481	417	370	400	439	-48.9	-44.0	-8.7
Alabama	—	—	—	—	—	—	—	26	5	6	21	12	10	4	9	-65.4	-19.2	-57.1
Arkansas	—	—	—	—	—	—	—	1	1	0	0	—	—	—	—	(1)	-100.0	(1)
Delaware	—	—	—	—	—	—	—	11	16	20	20	11	12	2	7	-36.4	81.8	-65.0
District of Columbia	—	—	—	—	—	—	—	1	0	1	2	—	—	1	—	(1)	100.0	(1)
Florida	—	—	—	—	—	—	—	123	117	63	46	36	37	32	26	-78.9	-62.6	-43.5
Georgia	—	—	—	—	—	—	—	4	1	2	7	8	4	6	9	125.0	75.0	28.6
Kentucky	—	—	—	—	—	—	—	15	14	22	16	6	16	6	3	-80.0	6.7	-81.3
Louisiana	—	—	—	—	—	—	—	0	1	3	9	1	2	0	5	(1)	(1)	-44.4
Maryland	—	—	—	—	—	—	—	12	8	8	11	11	11	10	15	25.0	-8.3	36.4
Mississippi	—	—	—	—	—	—	—	34	19	14	14	4	11	15	15	-55.9	-58.8	7.1
North Carolina	—	—	—	—	—	—	—	19	31	34	14	13	23	17	19	0.0	-26.3	35.7
Oklahoma	—	—	—	—	—	—	—	3	24	22	21	37	23	32	37	1133.3	600.0	76.2
South Carolina	—	—	—	—	—	—	—	0	0	0	0	6	2	6	1	(1)	(1)	(1)
Tennessee	—	—	—	—	—	—	—	7	22	22	10	26	46	37	33	371.4	42.9	230.0
Texas	—	—	—	—	—	—	—	191	171	133	116	133	79	118	121	-36.6	-39.3	4.3
Virginia	—	—	—	—	—	—	—	404	201	176	163	102	83	103	125	-69.1	-59.7	-23.3
West Virginia	—	—	—	—	—	—	—	8	9	10	11	11	11	11	14	75.0	37.5	27.3
West	—	—	—	—	—	—	—	600	615	580	556	564	613	583	692	15.3	-7.3	24.5
Alaska	—	—	—	—	—	—	—	7	3	1	0	—	—	—	—	(1)	-100.0	(1)
Arizona	—	—	—	—	—	—	—	20	20	5	9	3	1	3	4	-80.0	-55.0	-55.6
California	—	—	—	—	—	—	—	453	468	450	414	297	330	309	399	-11.9	-8.6	-3.6
Colorado	—	—	—	—	—	—	—	14	4	4	4	184	208	210	232	1557.1	-71.4	5700.0
Hawaii	—	—	—	—	—	—	—	0	0	0	0	—	—	—	—	(1)	(1)	(1)
Idaho	—	—	—	—	—	—	—	5	12	18	27	10	10	5	5	0.0	440.0	-81.5
Montana	—	—	—	—	—	—	—	0	1	0	0	—	1	—	—	(1)	(1)	(1)

Table 5-6.—Continued

Region and state	1975-76	1976-77	1977-78	1978-79	1979-80	1980-81	1981-82	1982-83	1983-84	1984-85	1985-86	1986-87	1987-88	1988-89	1989-90	Percent change: 1982-83 to 1989-90	Percent change: 1982-83 to 1985-86	Percent change: 1985-86 to 1989-90
West Continued																		
Nevada	—	—	—	—	—	—	—	6	2	8	2	2	2	6	—	(1)	-66.7	(1)
New Mexico	—	—	—	—	—	—	—	3	10	4	2	1	1	1	4	33.3	-33.3	100.0
Oregon	—	—	—	—	—	—	—	26	20	17	19	28	17	19	17	-34.6	-26.9	-10.5
Utah	—	—	—	—	—	—	—	0	0	0	0	3	—	2	10	(1)	(1)	(1)
Washington	—	—	—	—	—	—	—	47	57	59	65	25	31	17	13	-72.3	38.3	-80.0
Wyoming	—	—	—	—	—	—	—	19	18	14	14	11	12	11	8	-57.9	-26.3	-42.9
U.S. Service Schools	—	—	—	—	—	—	—	177	177	0	87	126	105	156	186	5.1	-50.8	113.8
Outlying Areas	—	—	—	—	—	—	—	35	35	36	31	17	26	11	16	-54.3	-11.4	-48.4
American Samoa	—	—	—	—	—	—	—	—	—	—	—	—	—	—	—	(1)	(1)	(1)
Guam	—	—	—	—	—	—	—	0	0	—	0	—	—	—	—	(1)	(1)	(1)
Northern Marianas	—	—	—	—	—	—	—	—	—	—	—	—	—	—	—	(1)	(1)	(1)
Puerto Rico	—	—	—	—	—	—	—	35	35	36	31	17	26	11	15	-57.1	-11.4	-51.6
Trust Territories	—	—	—	—	—	—	—	—	—	—	—	—	—	—	—	(1)	(1)	(1)
Virgin Islands	—	—	—	—	—	—	—	0	0	0	0	—	—	—	1	(1)	(1)	(1)

[1] Insufficient data for calculating a percent change.

—Data not available.

SOURCE: U.S. Department of Education, National Center for Education Statistics, HEGIS, "Degrees and Other Formal Awards Conferred" surveys; and IPEDS, "Completions" surveys.

Table 5-7.—Bachelor's degrees conferred in agricultural sciences, by region and state: 1975-76 to 1989-90

Region and state	1975-76	1976-77	1977-78	1978-79	1979-80	1980-81	1981-82	1982-83	1983-84	1984-85	1985-86	1986-87	1987-88	1988-89	1989-90	Percent change 1975-76 to 1989-90	1975-76 to 1980-81	1980-81 to 1985-86	1985-86 to 1989-90
United States	19,402	21,467	22,650	23,134	22,802	21,886	21,029	20,909	19,317	18,107	16,823	14,991	14,222	13,492	13,070	-32.6	12.8	-23.1	-22.3
Northeast	2,796	3,126	3,392	3,340	3,316	3,000	2,947	2,797	2,530	2,338	2,299	1,953	2,108	1,989	1,980	-29.2	7.3	-23.4	-13.9
Connecticut	219	240	293	263	214	184	139	124	108	109	92	83	67	40	49	-77.6	-16.0	-50.0	-46.7
Maine	177	165	229	228	269	213	180	219	144	140	135	151	152	110	112	-36.7	20.3	-36.6	-17.0
Massachusetts	378	392	437	419	359	307	293	242	239	203	169	164	134	114	119	-68.5	-18.8	-45.0	-29.6
New Hampshire	208	201	211	180	219	181	172	216	156	156	144	121	89	113	106	-49.0	-13.0	-20.4	-26.4
New Jersey	181	246	247	290	288	309	293	293	287	230	266	122	132	131	117	-35.4	70.7	-13.9	-56.0
New York	652	592	665	592	581	591	613	560	570	637	663	610	812	820	808	23.9	-9.4	12.2	21.9
Pennsylvania	662	861	876	946	996	835	911	832	736	599	569	480	493	432	464	-29.9	26.1	-31.9	-18.5
Rhode Island	162	209	223	211	186	179	144	109	104	77	77	78	72	76	59	-63.6	10.5	-57.0	-23.4
Vermont	157	220	211	211	204	201	202	202	186	187	184	144	157	153	146	-7.0	28.0	-8.5	-20.7
Midwest	6,208	6,781	7,100	7,201	7,319	7,238	6,990	7,116	6,730	6,193	5,616	5,161	4,736	4,730	4,343	-30.0	16.6	-22.4	-22.7
Illinois	734	836	862	893	893	858	848	871	828	780	665	610	595	585	523	-28.7	16.9	-22.5	-21.4
Indiana	535	594	624	599	660	607	545	578	470	484	431	387	307	302	279	-47.9	13.5	-29.0	-35.3
Iowa	515	500	578	610	628	712	645	579	599	554	531	481	415	410	378	-26.6	38.3	-25.4	-28.8
Kansas	439	457	468	473	460	453	490	424	432	412	410	353	343	357	280	-36.2	3.2	-9.5	-31.7
Michigan	939	1,062	1,185	1,087	1,126	1,167	1,113	1,066	913	749	622	650	645	683	631	-32.8	24.3	-46.7	1.4
Minnesota	383	416	401	437	426	402	340	408	355	334	276	233	208	172	172	-55.1	5.0	-31.3	-37.7
Missouri	679	730	770	806	740	794	787	845	830	803	729	670	585	521	458	-32.5	16.9	-8.2	-37.2
Nebraska	224	261	272	250	331	304	324	327	367	298	280	278	251	255	208	-7.1	35.7	-7.9	-25.7
North Dakota	178	202	206	205	236	226	190	190	187	200	172	164	152	139	147	-17.4	27.0	-23.9	-14.5
Ohio	734	837	791	840	698	671	603	639	569	519	465	393	363	379	332	-54.8	-8.6	-30.7	-28.6
South Dakota	189	160	202	198	227	181	205	248	228	219	214	172	170	158	166	-12.2	-4.2	18.2	-22.4
Wisconsin	659	726	741	803	894	863	900	941	952	841	821	770	702	769	769	16.7	31.0	-4.9	-6.3
South	5,708	6,478	6,845	7,177	6,846	6,645	6,209	5,968	5,695	5,559	5,111	4,584	4,288	3,933	4,060	-28.9	16.4	-23.1	-20.6
Alabama	281	291	333	344	374	367	335	309	281	219	212	185	198	167	178	-36.7	30.6	-42.2	-16.0
Arkansas	216	219	242	288	259	227	276	276	242	271	282	215	217	183	183	-15.3	5.1	24.2	-35.1
Delaware	85	90	105	181	166	166	148	148	123	112	112	101	108	89	93	9.4	95.3	-32.5	-17.0
District of Columbia	—	2	8	13	7	3	3	7	4	3	6	4	—	—	1	([1])	([1])	100.0	-83.3
Florida	278	353	410	398	412	316	310	314	252	254	267	230	191	222	188	-32.4	13.7	-15.5	-29.6
Georgia	287	392	444	459	454	445	367	349	361	375	285	282	238	218	225	-21.6	55.1	-36.0	-21.1
Kentucky	356	377	440	419	421	430	406	403	374	403	292	302	275	243	241	-32.3	20.8	-32.1	-17.5
Louisiana	489	528	493	480	476	408	394	343	366	363	318	262	322	266	252	-48.5	-16.6	-22.1	-20.8
Maryland	261	293	315	306	287	311	292	243	218	183	168	116	113	104	149	-42.9	19.2	-46.0	-11.3
Mississippi	250	314	262	256	235	250	208	200	204	204	195	174	164	157	139	-44.4	0.0	-22.0	-28.7
North Carolina	336	387	433	430	409	395	381	371	338	337	313	280	276	235	273	-18.8	17.6	-20.8	-12.8
Oklahoma	344	366	388	434	379	438	392	384	390	378	364	288	275	262	303	-11.9	27.3	-16.9	-16.8
South Carolina	161	155	184	188	183	153	156	162	132	124	112	106	87	106	101	-37.3	-5.0	-26.8	-9.8
Tennessee	424	538	544	575	551	532	484	404	437	409	370	308	326	282	216	-49.1	25.5	-30.5	-41.6
Texas	1,383	1,548	1,526	1,652	1,508	1,552	1,409	1,466	1,421	1,384	1,349	1,309	1,130	1,079	1,181	-14.6	12.2	-13.1	-12.5
Virginia	323	377	457	494	468	395	391	367	375	360	301	287	263	219	231	-28.5	22.3	-23.8	-23.3
West Virginia	234	248	261	260	257	257	257	222	177	180	165	135	105	101	106	-54.7	9.8	-35.8	-35.8
West	4,690	5,082	5,313	5,416	5,321	5,003	4,883	5,028	4,362	4,017	3,797	3,293	3,090	2,840	2,687	-42.7	6.7	-24.1	-29.2
Alaska	20	22	12	14	27	26	24	49	33	36	31	34	24	26	29	45.0	30.0	19.2	-6.5
Arizona	282	329	363	348	355	337	319	323	269	215	208	161	164	156	152	-46.1	19.5	-38.3	-26.9
California	1,742	1,907	2,145	2,196	2,222	2,038	2,027	2,166	2,041	1,868	1,741	1,531	1,471	1,315	1,285	-26.2	17.0	-14.6	-26.2
Colorado	448	479	428	520	486	484	449	535	403	423	424	369	306	302	304	-32.1	8.0	-12.4	-28.3
Hawaii	70	112	168	135	109	113	127	80	73	72	79	74	55	45	43	-38.6	61.4	-30.1	-45.6
Idaho	173	184	198	197	224	227	172	195	162	102	118	98	63	103	74	-57.2	31.2	-48.0	-37.3

Table 5-7.—Continued

Region and state	1975-76	1976-77	1977-78	1978-79	1979-80	1980-81	1981-82	1982-83	1983-84	1984-85	1985-86	1986-87	1987-88	1988-89	1989-90	Percent change 1975-76 to 1989-90	Percent change 1975-76 to 1980-81	Percent change 1980-81 to 1985-86	Percent change 1985-86 to 1989-90
West Continued																			
Montana	379	367	363	405	332	334	326	318	273	239	225	197	184	152	151	-60.2	-11.9	-32.6	-32.9
Nevada	69	54	65	64	71	62	51	55	35	35	44	37	34	36	23	-66.7	-10.1	-29.0	-47.7
New Mexico ..	188	208	257	240	223	192	207	161	142	125	136	117	129	99	96	-48.9	2.1	-29.2	-29.4
Oregon	459	448	348	333	394	349	348	407	268	300	259	231	212	226	203	-55.8	-24.0	-25.8	-21.6
Utah.........	318	335	352	345	325	307	318	244	234	212	174	152	159	121	106	-66.7	-3.5	-43.3	-39.1
Washington...	449	545	519	505	436	410	423	411	330	280	282	224	199	201	154	-65.7	-8.7	-31.2	-45.4
Wyoming	93	92	95	114	117	124	92	84	99	110	76	68	90	58	67	-28.0	33.3	-38.7	-11.8
U.S. Service Schools	—	—	—	—	—	—	—	—	—	—	—	—	—	—	—	([1])	([1])	([1])	([1])
Outlying Areas ..	58	64	74	113	101	147	125	115	83	76	63	91	109	72	96	65.5	153.4	-57.1	52.4
American Samoa.....	—	—	—	—	—	—	—	—	—	—	—	—	—	—	—	([1])	([1])	([1])	([1])
Guam	—	—	—	1	—	2	3	2	2	1	2	1	1	1	2	([1])	([1])	0.0	0.0
Northern Marianas ...	—	—	—	—	—	—	—	—	—	—	—	—	—	—	—	([1])	([1])	([1])	([1])
Puerto Rico ..	58	64	74	112	101	145	122	113	81	75	61	90	108	71	94	62.1	150.0	-57.9	54.1
Trust Territories	—	—	—	—	—	—	—	0	0	0	0	—	—	—	—	([1])	([1])	([1])	([1])
Virgin Islands .	—	—	—	—	—	—	—	0	0	0	0	—	—	—	—	([1])	([1])	([1])	([1])

[1] Insufficient data for calculating a percent change.

—Data not available.

SOURCE: U.S. Department of Education, National Center for Education Statistics, HEGIS, "Degrees and Other Formal Awards Conferred" surveys; and IPEDS, "Completions" surveys.

Table 5-8.—Bachelor's degrees conferred in computer sciences, by region and state: 1975-76 to 1989-90

Region and state	1975-76	1976-77	1977-78	1978-79	1979-80	1980-81	1981-82	1982-83	1983-84	1984-85	1985-86	1986-87	1987-88	1988-89	1989-90	Percent change: 1975-76 to 1989-90	Percent change: 1975-76 to 1980-81	Percent change: 1980-81 to 1985-86	Percent change: 1985-86 to 1989-90
United States	5,652	6,407	7,201	8,719	11,154	15,121	20,267	24,510	32,172	38,878	41,889	39,589	34,523	30,454	27,434	385.4	167.5	177.0	-34.5
Northeast	1,409	1,504	1,806	2,301	3,076	4,303	5,836	7,389	9,411	11,057	11,735	11,044	9,363	7,835	6,767	380.3	205.4	172.7	-42.3
Connecticut	67	61	111	100	138	144	179	205	211	250	373	348	309	305	280	317.9	114.9	159.0	-24.9
Maine	—	2	8	1	19	26	39	58	58	65	77	82	76	74	43	(1)	(1)	196.2	-44.2
Massachusetts	204	191	222	282	373	623	731	884	1,310	1,495	1,649	1,554	1,236	1,099	876	329.4	205.4	164.7	-46.9
New Hampshire	16	12	23	41	79	105	159	241	200	382	410	456	318	311	250	1462.5	556.3	290.5	-39.0
New Jersey	122	115	205	213	303	408	562	822	993	1,171	1,183	1,128	965	849	755	518.9	234.4	190.0	-36.2
New York	702	745	791	1,040	1,354	1,806	2,420	3,080	3,854	4,394	4,662	4,364	3,868	3,188	2,738	290.0	157.3	158.1	-41.3
Pennsylvania	263	310	390	513	698	1,029	1,515	1,865	2,495	2,963	2,924	2,638	2,233	1,731	1,579	500.4	291.3	184.2	-46.0
Rhode Island	35	62	48	92	96	142	203	199	259	282	375	377	288	243	196	460.0	305.7	164.1	-47.7
Vermont	—	6	8	19	16	20	28	35	31	55	82	97	70	35	50	(1)	(1)	310.0	-39.0
Midwest	1,642	2,030	2,185	2,531	3,156	4,301	5,720	6,650	8,626	10,667	10,760	10,134	8,847	8,161	7,237	340.7	161.9	150.2	-32.7
Illinois	298	467	499	591	697	951	1,344	1,557	1,990	2,295	2,244	2,173	2,000	1,689	1,560	423.5	219.1	136.0	-30.5
Indiana	149	153	169	184	299	301	380	509	783	980	977	937	801	707	640	329.5	102.0	224.6	-34.5
Iowa	115	135	123	135	160	281	369	387	526	648	703	581	499	432	324	181.7	144.3	150.2	-53.9
Kansas	88	107	119	130	160	185	247	317	357	455	443	398	341	318	301	242.0	110.2	139.5	-32.1
Michigan	268	293	335	408	467	737	770	998	1,220	1,471	1,554	1,384	1,248	1,122	976	264.2	175.0	110.9	-37.2
Minnesota	94	123	118	165	223	255	373	475	605	726	713	685	628	620	546	480.9	171.3	179.6	-23.4
Missouri	163	187	226	220	303	396	490	625	749	1,051	982	1,123	928	852	860	427.6	142.9	148.0	-12.4
Nebraska	33	41	41	50	40	69	97	107	159	205	240	255	223	201	152	360.6	109.1	247.8	-36.7
North Dakota	16	24	26	39	41	49	60	107	150	171	188	189	155	110	89	456.3	206.3	283.7	-52.7
Ohio	335	384	429	469	539	781	1,098	1,116	1,468	1,909	1,957	1,688	1,396	1,415	1,130	237.3	133.1	150.6	-42.3
South Dakota	19	37	25	35	67	51	59	112	122	172	171	130	133	108	81	326.3	168.4	235.3	-52.6
Wisconsin	64	79	75	105	160	245	433	340	497	584	588	591	495	587	578	803.1	282.8	140.0	-1.7
South	1,701	1,876	2,119	2,592	3,203	4,383	5,594	7,071	9,358	11,328	12,823	12,168	10,631	9,308	8,381	392.7	157.7	192.6	-34.6
Alabama	41	93	139	161	204	328	443	539	690	714	696	689	621	547	493	1102.4	700.0	112.2	-29.2
Arkansas	14	23	20	47	68	106	149	239	353	425	372	277	251	228	195	1292.9	657.1	250.9	-47.6
Delaware	29	20	25	30	32	46	48	72	64	70	69	64	69	74	74	155.2	58.6	50.0	7.2
District of Columbia	82	106	96	85	58	92	112	139	214	226	318	305	298	251	314	282.9	12.2	245.7	-1.3
Florida	259	295	274	379	407	600	705	922	1,154	1,560	1,672	1,465	1,308	1,070	989	281.9	131.7	178.7	-40.8
Georgia	61	84	87	114	136	163	225	432	700	906	1,009	1,081	904	793	782	1182.0	167.2	519.0	-22.5
Kentucky	45	60	70	92	129	164	206	274	419	445	417	402	327	296	245	444.4	264.4	154.3	-41.2
Louisiana	165	195	202	225	291	351	392	384	551	744	887	775	739	559	466	182.4	112.7	152.7	-47.5
Maryland	90	72	108	116	150	202	277	337	554	629	861	825	810	773	675	650.0	124.4	326.2	-21.6
Mississippi	117	91	84	134	152	191	228	300	312	458	504	471	347	294	208	77.8	63.2	163.9	-58.7
North Carolina	77	75	112	124	158	256	379	505	691	717	993	904	721	709	613	696.1	232.5	287.9	-38.3
Oklahoma	85	74	67	104	125	228	260	379	381	558	590	625	489	427	379	345.9	168.2	158.8	-35.8
South Carolina	49	41	46	58	63	85	106	159	247	348	433	400	348	400	209	326.5	73.5	409.4	-51.7
Tennessee	42	81	85	127	153	195	292	376	442	468	538	449	418	322	331	688.1	364.3	175.9	-38.5
Texas	411	443	513	638	802	1,055	1,394	1,413	1,686	1,927	2,202	2,167	1,823	1,535	1,496	264.0	156.7	108.7	-32.1
Virginia	95	94	136	124	224	250	291	495	734	926	1,053	1,063	985	877	768	708.4	163.2	321.2	-27.1
West Virginia	39	29	55	34	51	71	87	106	166	207	209	206	173	153	144	269.2	82.1	194.4	-31.1
West	841	960	1,047	1,269	1,700	2,094	3,071	3,354	4,680	5,749	6,430	6,224	5,604	5,056	4,966	490.5	149.0	207.1	-22.8
Alaska	1	—	1	2	—	—	1	3	7	17	17	16	23	27	25	2400.0	(1)	(1)	47.1
Arizona	49	64	107	122	147	179	321	328	468	740	712	496	421	355	360	634.7	265.3	297.8	-49.4
California	432	491	518	637	860	1,054	1,510	1,656	2,520	2,956	3,322	3,480	3,031	2,830	2,798	547.7	144.0	215.2	-15.8
Colorado	62	58	77	94	151	149	222	290	364	509	563	465	484	465	440	609.7	140.3	277.9	-21.8
Hawaii	8	21	24	44	73	81	122	86	79	106	137	150	153	132	139	1637.5	912.5	69.1	1.5
Idaho	10	10	10	15	14	38	50	91	134	151	165	134	116	78	76	660.0	280.0	334.2	-53.9

Table 5-8.—Continued

Region and state	1975-76	1976-77	1977-78	1978-79	1979-80	1980-81	1981-82	1982-83	1983-84	1984-85	1985-86	1986-87	1987-88	1988-89	1989-90	Percent change 1975-76 to 1989-90	Percent change 1975-76 to 1980-81	Percent change 1980-81 to 1985-86	Percent change 1985-86 to 1989-90
West Continued																			
Montana	28	23	15	21	30	44	75	57	93	133	125	125	118	87	72	157.1	57.1	184.1	-42.4
Nevada	—	—	—	—	2	20	13	18	23	32	36	48	28	24	36	(1)	(1)	80.0	0.0
New Mexico	52	70	49	48	67	66	103	115	111	129	169	206	170	162	158	203.8	26.9	156.1	-6.5
Oregon	67	78	81	91	113	132	150	213	249	295	368	268	291	244	271	304.5	97.0	178.8	-26.4
Utah	90	94	91	100	142	165	192	211	310	298	413	425	353	303	288	220.0	83.3	150.3	-30.3
Washington	34	45	64	80	88	152	283	276	290	338	344	356	366	308	262	670.6	347.1	126.3	-23.8
Wyoming	8	6	10	15	13	14	29	10	32	45	59	55	50	41	41	412.5	75.0	321.4	-30.5
U.S. Service Schools	59	37	44	26	19	40	46	46	97	77	141	19	78	94	83	40.7	-32.2	252.5	-41.1
Outlying Areas	12	19	23	50	59	112	164	172	263	243	306	338	348	326	261	2075.0	833.3	173.2	-14.7
American Samoa	—	—	—	—	—	—	—	—	—	—	—	—	—	—	—	(1)	(1)	(1)	(1)
Guam	—	—	—	—	—	—	—	0	0	0	0	—	—	—	—	(1)	(1)	(1)	(1)
Northern Marianas	—	—	—	—	—	—	—	—	—	—	—	—	—	—	—	(1)	(1)	(1)	(1)
Puerto Rico	12	19	23	50	59	112	164	172	263	243	306	338	348	326	261	2075.0	833.3	173.2	-14.7
Trust Territories	—	—	—	—	—	—	—	—	—	—	—	—	—	—	—	(1)	(1)	(1)	(1)
Virgin Islands	—	—	—	—	—	—	—	—	—	0	—	—	—	—	—	(1)	(1)	(1)	(1)

[1] Insufficient data for calculating a percent change.

—Data not available.

SOURCE: U.S. Department of Education, National Center for Education Statistics, HEGIS, "Degrees and Other Formal Awards Conferred" surveys; and IPEDS, "Completions" surveys.

Table 5-9.—Bachelor's degrees conferred in engineering, by region and state: 1975-76 to 1989-90

Region and state	1975-76	1976-77	1977-78	1978-79	1979-80	1980-81	1981-82	1982-83	1983-84	1984-85	1985-86	1986-87	1987-88	1988-89	1989-90	Percent change: 1975-76 to 1989-90	Percent change: 1975-76 to 1980-81	Percent change: 1980-81 to 1985-86	Percent change: 1985-86 to 1989-90
United States . . .	46,331	49,283	55,654	62,375	68,893	75,000	80,005	89,270	94,444	96,105	95,953	93,096	88,706	85,225	82,110	77.2	61.9	27.9	-14.4
Northeast	11,787	12,577	14,116	15,741	16,644	17,442	18,748	20,654	22,198	22,199	22,070	21,376	20,109	19,226	18,487	56.8	48.0	26.5	-16.2
Connecticut . .	455	606	595	635	698	758	824	984	1,048	1,153	1,159	1,081	987	850	855	87.9	66.6	52.9	-26.2
Maine	247	252	322	341	328	321	323	451	478	436	430	429	349	342	306	23.9	30.0	34.0	-28.8
Massachusetts	2,483	2,504	2,885	3,075	3,182	3,252	3,806	4,068	4,563	4,595	4,730	4,695	4,433	4,254	3,876	56.1	31.0	45.4	-18.1
New Hampshire	133	168	225	243	284	313	365	396	393	357	379	331	323	346	353	165.4	135.3	21.1	-6.9
New Jersey . . .	1,324	1,472	1,472	1,645	1,630	1,772	1,962	2,125	2,112	2,185	2,201	2,129	2,016	1,739	1,708	29.0	33.8	24.2	-22.4
New York	4,069	4,174	4,779	5,219	5,603	5,663	5,878	6,336	7,177	6,915	6,750	6,634	6,119	6,229	5,939	46.0	39.2	19.2	-12.0
Pennsylvania .	2,675	2,998	3,410	4,063	4,376	4,726	4,960	5,634	5,679	5,861	5,697	5,398	5,181	4,817	4,859	81.6	76.7	20.5	-14.7
Rhode Island .	274	287	315	365	365	452	439	486	522	492	484	485	509	465	373	36.1	65.0	7.1	-22.9
Vermont	127	116	113	155	178	185	191	174	226	205	240	194	192	184	218	71.7	45.7	29.7	-9.2
Midwest	13,195	14,286	15,547	17,616	19,467	21,570	23,288	26,179	27,474	27,893	27,658	26,716	25,479	25,373	24,617	86.6	63.5	28.2	-11.0
Illinois	2,300	2,418	2,617	2,939	3,415	3,692	3,994	4,101	4,366	4,358	4,595	4,534	4,610	4,464	4,293	86.7	60.5	24.5	-6.6
Indiana	1,746	1,899	2,017	2,320	2,773	2,888	3,216	3,501	3,580	3,568	3,601	3,383	3,187	3,269	3,225	84.7	65.4	24.7	-10.4
Iowa	522	566	588	757	777	990	878	1,038	1,079	1,234	1,289	1,172	1,205	1,128	1,017	94.8	89.7	30.2	-21.1
Kansas	609	702	794	825	883	908	988	1,146	1,256	1,047	1,045	962	910	963	913	49.9	49.1	15.1	-12.6
Michigan	2,740	2,758	3,115	3,427	3,765	4,214	4,461	4,924	5,124	4,984	4,897	4,831	4,778	4,599	4,739	73.0	53.8	16.2	-3.2
Minnesota	538	642	607	765	801	908	990	1,145	1,238	1,202	1,059	1,073	1,094	1,079	1,058	96.7	68.8	16.6	-0.1
Missouri	1,127	1,230	1,314	1,539	1,739	1,806	2,039	2,358	2,486	2,644	2,439	2,212	2,096	1,948	1,906	69.1	60.2	35.0	-21.9
Nebraska	222	241	253	237	286	335	324	418	432	445	448	410	355	351	334	50.5	50.9	33.7	-25.4
North Dakota .	166	171	210	258	270	311	400	500	505	631	588	486	470	472	433	160.8	87.3	89.1	-26.4
Ohio	1,945	2,189	2,377	2,742	2,961	3,388	3,584	4,303	4,456	4,788	4,780	4,710	4,241	4,237	3,954	103.3	74.2	41.1	-17.3
South Dakota .	253	332	355	335	382	400	496	509	577	567	526	503	397	402	353	39.5	58.1	31.5	-32.9
Wisconsin	1,027	1,138	1,300	1,472	1,415	1,730	1,918	2,236	2,375	2,425	2,391	2,440	2,136	2,461	2,392	132.9	68.5	38.2	0.0
South	12,552	12,980	15,484	17,442	19,531	21,795	22,583	25,226	26,582	27,368	27,395	25,982	24,774	23,214	22,383	78.3	73.6	25.7	-18.3
Alabama	604	674	778	909	1,123	1,194	1,368	1,370	1,529	1,666	1,600	1,703	1,672	1,618	1,534	154.0	97.7	34.0	-4.1
Arkansas	184	141	150	189	287	323	345	359	407	436	391	330	323	372	330	79.3	75.5	21.1	-15.6
Delaware	181	214	236	272	263	282	260	291	288	265	261	232	237	206	204	12.7	55.8	-7.4	-21.8
District of Columbia . . .	187	144	196	255	312	379	388	449	458	446	449	461	427	446	357	90.9	102.7	18.5	-20.5
Florida	1,206	1,302	1,513	1,614	1,434	1,868	1,830	2,784	3,012	3,398	3,599	3,289	3,513	3,416	3,472	187.9	54.9	92.7	-3.5
Georgia	1,059	1,046	1,203	1,335	1,459	1,659	1,784	2,012	1,973	1,904	1,789	1,857	1,796	1,726	1,907	80.1	56.7	7.8	6.6
Kentucky	475	481	536	589	782	787	840	944	925	940	917	825	868	767	763	60.6	65.7	16.5	-16.8
Louisiana	891	895	1,020	1,225	1,446	1,577	1,704	1,908	1,932	2,062	1,921	1,875	1,660	1,455	1,210	35.8	77.0	21.8	-37.0
Maryland	401	428	514	549	721	835	802	1,018	951	1,012	1,099	1,125	1,038	1,056	993	147.6	108.2	31.6	-9.6
Mississippi . . .	353	313	368	507	573	677	646	697	721	679	760	803	633	585	600	70.0	91.8	12.3	-21.1
North Carolina	877	900	1,053	1,046	1,226	1,274	1,451	1,523	1,752	1,742	1,751	1,641	1,592	1,478	1,615	84.2	45.3	37.4	-7.8
Oklahoma	737	713	871	1,043	1,085	1,170	1,152	1,316	1,326	1,390	1,349	1,259	1,042	1,032	977	32.6	58.8	15.3	-27.6
South Carolina	402	567	725	722	825	907	730	821	882	960	987	908	766	908	695	72.9	125.6	8.8	-29.6
Tennessee . . .	952	935	1,096	1,328	1,449	1,603	1,782	1,797	1,876	1,980	2,066	1,974	1,862	1,668	1,529	60.6	68.4	28.9	-26.0
Texas	2,838	2,941	3,541	3,976	4,429	4,944	5,032	5,253	5,705	5,604	5,665	5,207	4,855	4,271	4,098	44.4	74.2	14.6	-27.7
Virginia	869	910	1,160	1,338	1,475	1,565	1,727	1,842	1,960	2,017	1,904	1,789	1,942	1,719	1,569	80.6	80.1	21.7	-17.6
West Virginia .	336	376	524	545	642	751	742	842	885	867	887	704	548	491	530	57.7	123.5	18.1	-40.2
West	8,098	8,715	9,593	10,627	12,398	13,285	14,389	16,223	16,931	17,436	17,775	18,026	17,472	16,548	15,829	95.5	64.1	33.8	-10.9
Alaska	24	22	36	29	34	38	50	64	66	83	115	120	95	90	62	158.3	58.3	202.6	-46.1
Arizona	716	786	765	870	1,127	1,160	1,286	1,348	1,585	1,536	1,606	1,706	1,590	1,582	1,364	90.5	62.0	38.4	-15.1
California	3,679	3,849	4,308	4,801	5,591	6,127	6,704	7,648	7,888	8,006	8,390	8,839	8,500	8,067	8,091	119.9	66.5	36.9	-3.6
Colorado	935	1,073	1,084	1,195	1,446	1,566	1,711	1,798	1,768	1,816	1,820	1,701	1,663	1,627	1,543	65.0	67.5	16.2	-15.2
Hawaii	141	128	174	177	211	174	192	159	204	212	237	212	254	184	183	29.8	23.4	36.2	-22.8
Idaho	109	125	133	164	213	188	235	239	264	288	277	247	235	197	197	80.7	72.5	47.3	-28.9

Table 5-9.—Continued

Region and state	1975-76	1976-77	1977-78	1978-79	1979-80	1980-81	1981-82	1982-83	1983-84	1984-85	1985-86	1986-87	1987-88	1988-89	1989-90	Percent change 1975-76 to 1989-90	Percent change 1975-76 to 1980-81	Percent change 1980-81 to 1985-86	Percent change 1985-86 to 1989-90
West Continued																			
Montana	251	253	306	347	418	423	534	590	657	735	658	589	561	462	447	78.1	68.5	55.6	-32.1
Nevada	73	65	80	85	103	97	110	134	153	171	138	155	155	147	142	94.5	32.9	42.3	2.9
New Mexico	351	352	390	446	547	547	516	658	678	745	726	680	691	592	526	49.9	55.8	32.7	-27.5
Oregon	498	559	616	638	703	778	851	893	874	825	939	905	1,013	931	900	80.7	56.2	20.7	-4.2
Utah	506	602	637	700	722	857	826	1,132	1,175	1,299	1,256	1,214	1,113	1,077	940	85.8	69.4	46.6	-25.2
Washington	707	797	913	1,012	1,124	1,164	1,188	1,363	1,388	1,490	1,374	1,412	1,413	1,409	1,296	83.3	64.6	18.0	-5.7
Wyoming	108	104	151	163	159	166	186	197	231	230	239	246	189	183	138	27.8	53.7	44.0	-42.3
U.S. Service Schools	699	725	914	949	853	908	997	988	1,259	1,209	1,055	996	872	864	794	13.6	29.9	16.2	-24.7
Outlying Areas	386	394	355	425	372	395	391	484	461	458	545	679	723	759	744	92.7	2.3	38.0	36.5
American Samoa	—	—	—	—	—	—	—	—	—	—	—	—	—	—	—	([1])	([1])	([1])	([1])
Guam	—	1	—	1	—	1	1	0	0	0	0	—	—	—	—	([1])	([1])	-100.0	([1])
Northern Marianas	—	—	—	—	—	—	—	—	—	—	—	—	—	—	—	([1])	([1])	([1])	([1])
Puerto Rico	386	392	354	424	372	394	390	484	461	458	545	679	723	759	744	92.7	2.1	38.3	36.5
Trust Territories	—	—	—	—	—	—	—	—	—	—	—	—	—	—	—	([1])	([1])	([1])	([1])
Virgin Islands	—	1	1	—	—	—	—	—	—	—	—	—	—	—	—	([1])	([1])	([1])	([1])

[1] Insufficient data for calculating a percent change.

—Data not available.

SOURCE: U.S. Department of Education, National Center for Education Statistics, HEGIS, "Degrees and Other Formal Awards Conferred" surveys; and IPEDS, "Completions" surveys.

Table 5-10.—Bachelor's degrees conferred in health sciences, by region and state: 1975-76 to 1989-90

Region and state	1975-76	1976-77	1977-78	1978-79	1979-80	1980-81	1981-82	1982-83	1983-84	1984-85	1985-86	1986-87	1987-88	1988-89	1989-90	Percent change			
																1975-76 to 1989-90	1975-76 to 1980-81	1980-81 to 1985-86	1985-86 to 1989-90
United States	53,813	57,122	59,168	61,819	63,607	63,348	63,385	64,614	64,338	64,513	64,535	63,213	60,754	59,138	58,816	9.3	17.7	1.9	-8.9
Northeast	13,232	14,448	14,924	15,614	16,111	16,722	16,769	16,776	16,806	16,461	16,492	16,163	15,797	14,977	14,567	10.1	26.4	-1.4	-11.7
Connecticut	700	820	838	876	862	881	824	854	804	830	871	774	802	699	694	-0.9	25.9	-1.1	-20.3
Maine	170	154	133	139	394	466	300	433	545	599	671	727	765	656	669	293.5	174.1	44.0	-0.3
Massachusetts	2,090	2,529	2,539	2,231	2,380	2,516	2,636	2,743	2,869	2,647	2,492	2,554	2,529	2,345	2,333	11.6	20.4	-1.0	-6.4
New Hampshire	220	247	243	262	284	292	300	331	293	317	297	244	263	233	215	-2.3	32.7	1.7	-27.6
New Jersey	1,086	1,243	1,276	1,247	1,245	1,381	1,332	1,396	1,238	1,186	1,154	1,164	1,043	1,031	985	-9.3	27.2	-16.4	-14.6
New York	5,434	5,600	5,982	6,612	6,623	6,685	6,662	6,066	6,013	5,692	5,690	5,418	5,167	5,142	4,836	-11.0	23.0	-14.9	-15.0
Pennsylvania	2,879	3,128	3,223	3,468	3,595	3,773	3,895	4,240	4,333	4,440	4,594	4,586	4,570	4,313	4,262	48.0	31.1	21.8	-7.2
Rhode Island	482	554	529	606	541	554	617	525	519	514	521	486	449	380	380	-21.2	14.9	-6.0	-27.1
Vermont	171	173	161	173	187	174	203	188	192	236	202	210	209	178	193	12.9	1.8	16.1	-4.5
Midwest	16,295	17,086	17,682	18,890	19,160	19,503	19,468	19,903	19,724	20,073	20,266	20,327	19,206	19,362	19,232	18.0	19.7	3.9	-5.1
Illinois	2,471	3,013	3,122	3,593	3,748	3,954	3,720	3,659	3,755	3,747	3,876	3,877	3,587	3,604	3,526	42.7	60.0	-2.0	-9.0
Indiana	1,693	1,723	1,707	1,647	1,661	1,648	1,683	2,041	1,849	1,939	2,000	2,117	2,014	1,876	1,894	11.9	-2.7	21.4	-5.3
Iowa	735	721	768	924	934	894	880	909	826	821	916	916	818	795	768	4.5	21.6	2.5	-16.2
Kansas	788	845	875	1,085	1,022	1,030	1,116	1,191	1,140	1,108	1,064	937	856	831	977	24.0	30.7	3.3	-8.2
Michigan	2,380	2,472	2,404	2,768	2,801	2,862	3,046	2,945	2,985	3,034	3,083	3,018	2,855	3,067	2,986	25.5	20.3	7.7	-3.1
Minnesota	1,295	1,169	1,182	1,229	1,225	1,215	1,305	1,282	1,301	1,273	1,164	1,218	1,153	1,052	1,016	-21.5	-6.2	-4.2	-12.7
Missouri	1,173	1,338	1,427	1,483	1,461	1,592	1,532	1,594	1,542	1,796	1,555	1,567	1,487	1,668	1,663	41.8	35.7	-2.3	6.9
Nebraska	690	664	850	706	625	566	635	611	628	684	708	683	726	595	669	-3.0	-18.0	25.1	-5.5
North Dakota	522	569	517	549	562	541	506	456	461	394	488	501	496	526	489	-6.3	3.6	-9.8	0.2
Ohio	2,275	2,326	2,339	2,460	2,530	2,654	2,520	2,644	2,737	2,840	2,981	2,977	2,950	3,044	2,892	27.1	16.7	12.3	-3.0
South Dakota	392	405	365	360	344	342	410	395	380	419	368	348	307	335	354	-9.7	-12.8	7.6	-3.8
Wisconsin	1,881	1,841	2,126	2,086	2,247	2,205	2,115	2,176	2,120	2,018	2,063	2,168	1,957	1,969	1,998	6.2	17.2	-6.4	-3.2
South	16,025	17,337	17,723	17,999	18,153	17,851	17,907	18,400	18,967	19,050	18,894	18,165	17,466	17,301	17,314	8.0	11.4	5.8	-8.4
Alabama	1,172	1,171	1,214	1,351	1,354	1,373	1,292	1,436	1,458	1,472	1,396	1,313	1,286	1,292	1,269	8.3	17.2	1.7	-9.1
Arkansas	262	390	409	432	446	430	449	452	558	491	501	463	519	439	516	96.9	64.1	16.5	3.0
Delaware	211	236	268	270	244	211	257	267	234	273	240	243	224	199	198	-6.2	0.0	13.7	-17.5
District of Columbia	474	539	614	614	564	589	565	553	558	600	579	549	457	500	451	-4.9	24.3	-1.7	-22.1
Florida	1,029	1,275	1,244	1,479	1,376	1,446	1,325	1,418	1,615	1,823	1,641	1,613	1,453	1,515	1,615	56.9	40.5	13.5	-1.6
Georgia	1,260	1,267	1,230	1,092	1,257	1,170	1,173	1,199	1,031	1,081	1,047	1,070	1,087	988	1,120	-11.1	-7.1	-10.5	7.0
Kentucky	634	666	770	805	734	678	751	720	791	821	908	905	926	899	944	48.9	6.9	33.9	4.0
Louisiana	1,201	1,228	1,235	1,120	1,183	1,073	1,135	1,316	1,297	1,225	1,352	1,273	1,255	1,280	1,267	5.5	-10.7	26.0	-6.3
Maryland	1,113	1,082	1,168	1,058	1,079	1,071	1,159	1,176	1,165	1,083	1,101	1,032	1,045	1,031	995	-10.6	-3.8	2.8	-9.6
Mississippi	519	632	587	617	664	620	600	664	729	663	678	715	619	530	576	11.0	19.5	9.4	-15.0
North Carolina	1,248	1,456	1,449	1,437	1,514	1,573	1,550	1,550	1,544	1,568	1,497	1,474	1,437	1,353	1,303	4.4	26.0	-4.8	-13.0
Oklahoma	854	855	853	834	855	878	700	687	842	866	886	842	799	862	913	6.9	2.8	0.9	3.0
South Carolina	574	604	615	503	559	499	602	561	649	629	697	765	701	765	690	20.2	-13.1	39.7	-1.0
Tennessee	877	981	966	910	939	927	1,013	983	1,026	1,057	1,086	914	872	852	884	0.8	5.7	17.2	-18.6
Texas	3,170	3,425	3,443	3,674	3,696	3,506	3,484	3,547	3,474	3,428	3,326	3,080	3,052	2,973	2,896	-8.6	10.6	-5.1	-12.9
Virginia	974	1,036	1,085	1,218	1,132	1,212	1,263	1,242	1,334	1,329	1,339	1,303	1,168	1,238	1,133	16.3	24.4	10.5	-15.4
West Virginia	453	494	573	585	557	595	589	629	662	641	620	611	566	585	544	20.1	31.3	4.2	-12.3
West	8,261	8,251	8,839	9,316	10,183	9,272	9,241	9,535	8,841	8,929	8,883	8,558	8,285	7,498	7,703	-6.8	12.2	-4.2	-13.3
Alaska	39	23	23	22	35	29	50	54	41	48	52	40	61	49	42	7.7	-25.6	79.3	-19.2
Arizona	561	620	517	521	621	631	701	899	665	872	796	800	744	510	514	-8.4	12.5	26.1	-35.4
California	3,531	3,551	4,167	4,515	5,004	4,630	4,424	4,626	4,343	4,270	4,117	3,751	3,649	3,283	3,478	-1.5	31.1	-11.1	-15.5
Colorado	818	804	755	775	979	840	856	822	794	778	780	806	744	761	773	-5.5	2.7	-7.1	-0.9
Hawaii	125	130	106	106	131	130	110	136	154	86	98	135	138	121	120	-4.0	4.0	-24.6	22.4
Idaho	212	243	230	210	256	199	211	183	178	157	158	154	159	166	209	-1.4	-6.1	-20.6	32.3

Table 5-10.—Continued

Region and state	1975-76	1976-77	1977-78	1978-79	1979-80	1980-81	1981-82	1982-83	1983-84	1984-85	1985-86	1986-87	1987-88	1988-89	1989-90	Percent change 1975-76 to 1989-90	Percent change 1975-76 to 1980-81	Percent change 1980-81 to 1985-86	Percent change 1985-86 to 1989-90
West Continued																			
Montana	285	319	328	341	283	269	266	255	295	289	286	308	281	269	281	-1.4	-5.6	6.3	-1.7
Nevada	121	107	97	96	132	83	91	132	126	116	136	90	110	113	122	0.8	-31.4	63.9	-10.3
New Mexico ..	203	186	217	224	254	195	215	210	217	228	250	243	229	205	249	22.7	-3.9	28.2	-0.4
Oregon	496	583	620	638	618	510	588	539	443	493	577	596	574	550	547	10.3	2.8	13.1	-5.2
Utah.........	385	443	411	544	536	421	391	354	353	329	428	417	443	429	412	7.0	9.4	1.7	-3.7
Washington...	1,346	1,076	1,225	1,212	1,223	1,220	1,232	1,233	1,130	1,161	1,092	1,120	1,056	930	834	-38.0	-9.4	-10.5	-23.6
Wyoming.....	139	166	143	112	111	115	106	92	102	102	113	98	97	112	122	-12.2	-17.3	-1.7	8.0
U.S. Service Schools......	—	—	—	—	—	—	—	0	0	0	0	—	—	—	—	([1])	([1])	([1])	([1])
Outlying Areas..	437	517	593	637	667	1,021	941	1,056	1,039	1,099	1,197	1,186	945	1,031	1,036	137.1	133.6	17.2	-13.5
American Samoa.....	—	—	—	—	—	—	—	0	0	0	0	—	—	—	—	([1])	([1])	([1])	([1])
Guam	—	—	—	—	—	—	—	0	0	2	1	1	1	6	6	([1])	([1])	([1])	500.0
Northern Marianas...	22	8	3	6	—	—	—	—	—	—	—	—	—	—	—	([1])	([1])	([1])	([1])
Puerto Rico ..	415	509	590	631	667	1,021	941	1,056	1,039	1,097	1,196	1,175	936	1,021	1,023	146.5	146.0	17.1	-14.5
Trust Territories ..	—	—	—	—	—	—	—	—	—	—	—	—	—	—	—	([1])	([1])	([1])	([1])
Virgin Islands .	—	—	—	—	—	—	—	0	0	0	—	10	8	4	7	([1])	([1])	([1])	([1])

[1] Insufficient data for calculating a percent change.

—Data not available.

SOURCE: U.S. Department of Education, National Center for Education Statistics, HEGIS, "Degrees and Other Formal Awards Conferred" surveys; and IPEDS, "Completions" surveys.

Table 5-11.—Bachelor's degrees conferred in life sciences, by region and state: 1975-76 to 1989-90

Region and state	1975-76	1976-77	1977-78	1978-79	1979-80	1980-81	1981-82	1982-83	1983-84	1984-85	1985-86	1986-87	1987-88	1988-89	1989-90	Percent change: 1975-76 to 1989-90	Percent change: 1975-76 to 1980-81	Percent change: 1980-81 to 1985-86	Percent change: 1985-86 to 1989-90
United States . . .	54,275	53,605	51,502	48,846	46,370	43,216	41,639	39,982	38,640	38,445	38,524	38,121	36,755	36,059	37,170	-31.5	-20.4	-10.9	-3.5
Northeast	14,909	14,624	14,505	13,695	13,326	12,317	11,720	11,116	10,536	10,488	10,480	10,096	9,464	9,227	8,939	-40.0	-17.4	-14.9	-14.7
Connecticut . .	946	880	895	835	759	800	648	630	629	630	598	539	534	491	497	-47.5	-15.4	-25.3	-16.9
Maine	366	366	305	335	349	285	367	287	232	210	231	253	223	217	186	-49.2	-22.1	-18.9	-19.5
Massachusetts	2,168	2,195	2,264	2,287	2,116	2,018	1,888	1,806	1,908	1,854	1,948	1,804	1,686	1,521	1,477	-31.9	-6.9	-3.5	-24.2
New Hampshire	380	353	347	341	315	275	269	312	238	251	243	205	210	208	206	-45.8	-27.6	-11.6	-15.2
New Jersey . . .	1,663	1,701	1,777	1,542	1,512	1,491	1,319	1,337	1,125	1,114	1,039	1,048	968	1,059	939	-43.5	-10.3	-30.3	-9.6
New York	5,190	5,200	5,072	4,616	4,587	4,034	3,977	3,684	3,346	3,450	3,421	3,341	3,095	3,032	2,898	-44.2	-22.3	-15.2	-15.3
Pennsylvania .	3,466	3,231	3,190	3,079	3,047	2,847	2,614	2,555	2,546	2,466	2,487	2,459	2,360	2,247	2,304	-33.5	-17.9	-12.6	-7.4
Rhode Island .	439	405	399	405	375	381	418	354	371	314	316	296	243	271	254	-42.1	-13.2	-17.1	-19.6
Vermont	291	293	256	255	266	186	220	151	141	199	197	151	145	181	178	-38.8	-36.1	5.9	-9.6
Midwest	13,884	13,712	12,980	12,201	11,580	10,888	10,470	9,960	9,889	9,958	9,839	9,967	9,647	9,327	9,394	-32.3	-21.6	-9.6	-4.5
Illinois	2,772	2,847	2,596	2,575	2,434	2,390	2,157	1,995	2,067	2,065	2,053	2,065	2,018	2,004	1,989	-28.2	-13.8	-14.1	-3.1
Indiana	1,329	1,293	1,255	1,134	964	972	906	928	866	935	878	895	820	762	825	-37.9	-26.9	-9.7	-6.0
Iowa	734	680	668	663	620	553	581	581	567	576	583	568	606	586	577	-21.4	-24.7	5.4	-1.0
Kansas	564	624	564	553	550	489	564	482	510	512	464	459	432	419	456	-19.1	-13.3	-5.1	-1.7
Michigan	2,280	2,260	2,142	1,903	1,770	1,691	1,564	1,529	1,523	1,545	1,468	1,452	1,417	1,353	1,293	-43.3	-25.8	-13.2	-11.9
Minnesota	1,053	1,000	980	946	1,032	866	874	829	836	866	774	856	815	749	792	-24.8	-17.8	-10.6	2.3
Missouri	1,074	1,062	979	955	821	734	783	788	685	703	704	706	742	669	698	-35.0	-31.7	-4.1	-0.9
Nebraska	454	387	438	402	404	373	375	370	319	351	348	347	334	293	272	-40.1	-17.8	-6.7	-21.8
North Dakota .	123	110	109	108	89	95	66	92	125	76	99	103	88	99	89	-27.6	-22.8	4.2	-10.1
Ohio	2,262	2,189	2,026	1,840	1,739	1,626	1,586	1,422	1,411	1,384	1,438	1,424	1,344	1,358	1,386	-38.7	-28.1	-11.6	-3.6
South Dakota .	159	173	196	144	175	154	131	150	148	125	136	158	139	132	122	-23.3	-3.1	-11.7	-10.3
Wisconsin	1,080	1,087	1,027	978	982	945	883	794	832	820	894	934	892	903	895	-17.1	-12.5	-5.4	0.1
South	14,520	14,570	14,022	13,386	12,774	11,720	11,370	10,675	10,395	10,228	10,308	10,229	10,082	9,970	10,546	-27.4	-19.3	-12.0	2.3
Alabama	712	747	767	729	626	629	615	518	462	482	474	459	461	461	462	-35.1	-11.7	-24.6	-2.5
Arkansas	358	374	309	305	287	306	282	298	288	234	245	253	230	227	249	-30.4	-14.5	-19.9	1.6
Delaware	223	235	207	136	127	115	104	108	137	134	122	135	141	110	106	-52.5	-48.4	6.1	-13.1
District of Columbia . . .	358	353	383	340	329	303	300	334	294	264	266	275	256	259	250	-30.2	-15.4	-12.2	-6.0
Florida	1,141	1,169	1,077	983	941	887	861	805	793	720	819	832	807	855	888	-22.2	-22.3	-7.7	8.4
Georgia	816	780	801	844	772	741	744	606	637	681	641	632	619	564	656	-19.6	-9.2	-13.5	2.3
Kentucky	606	654	609	553	604	530	487	438	492	438	476	460	427	430	427	-29.5	-12.5	-10.2	-10.3
Louisiana	695	729	727	647	551	574	592	485	451	444	426	438	432	403	438	-37.0	-17.4	-25.8	2.8
Maryland	918	906	859	827	812	731	698	730	727	692	688	726	759	717	771	-16.0	-20.4	-5.9	12.1
Mississippi . . .	425	439	416	368	358	360	321	341	299	284	348	332	330	340	310	-27.1	-15.3	-3.3	-10.9
North Carolina	1,301	1,249	1,322	1,247	1,213	1,197	1,131	1,025	1,064	1,042	1,013	1,045	1,067	1,085	1,159	-10.9	-8.0	-15.4	14.4
Oklahoma	616	594	533	526	497	442	437	423	433	421	390	392	362	376	379	-38.5	-28.2	-11.8	-2.8
South Carolina	692	744	695	661	681	551	557	548	518	550	541	501	531	500	605	-12.6	-20.4	-1.8	11.8
Tennessee . . .	1,142	1,092	1,001	986	936	869	818	751	755	726	695	698	588	568	640	-44.0	-23.9	-20.0	-7.9
Texas	2,817	2,843	2,662	2,621	2,572	2,093	2,000	1,934	1,755	1,830	1,888	1,879	1,837	1,978	1,989	-29.4	-25.7	-9.8	5.3
Virginia	1,366	1,322	1,347	1,349	1,222	1,170	1,202	1,114	1,088	1,100	1,100	1,007	1,063	916	1,042	-23.7	-14.3	-6.0	-5.3
West Virginia .	334	340	307	264	246	222	221	217	202	186	176	165	172	181	175	-47.6	-33.5	-20.7	-0.6
West	10,878	10,633	9,972	9,527	8,629	8,241	8,053	8,191	7,794	7,742	7,868	7,806	7,523	7,497	8,256	-24.1	-24.2	-4.5	4.9
Alaska	41	44	33	20	40	42	33	49	41	44	36	34	35	29	35	-14.6	2.4	-14.3	-2.8
Arizona	460	468	466	450	349	339	298	353	302	309	306	330	294	312	347	-24.6	-26.3	-9.7	13.4
California	6,590	6,477	6,149	5,828	5,218	4,995	4,965	4,936	4,703	4,690	4,830	4,816	4,677	4,584	5,184	-21.3	-24.2	-3.3	7.3
Colorado	920	925	947	837	753	716	647	679	695	639	651	633	628	647	752	-18.3	-22.2	-9.1	15.5
Hawaii	154	151	167	137	115	106	113	114	145	135	143	128	89	118	110	-28.6	-31.2	34.9	-23.1
Idaho	126	128	131	134	140	89	114	121	110	113	103	105	105	104	111	-11.9	-29.4	15.7	7.8

Table 5-11.—Continued

Region and state	1975-76	1976-77	1977-78	1978-79	1979-80	1980-81	1981-82	1982-83	1983-84	1984-85	1985-86	1986-87	1987-88	1988-89	1989-90	Percent change 1975-76 to 1989-90	Percent change 1975-76 to 1980-81	Percent change 1980-81 to 1985-86	Percent change 1985-86 to 1989-90
West Continued																			
Montana	261	200	186	191	188	186	165	176	193	180	180	176	191	131	155	-40.6	-28.7	-3.2	-13.9
Nevada	57	52	69	41	47	58	48	46	58	50	55	58	61	49	59	3.5	1.8	-5.2	7.3
New Mexico	220	192	189	214	180	163	129	143	107	136	139	137	131	157	149	-32.3	-25.9	-14.7	7.2
Oregon	528	499	492	476	409	398	429	467	391	388	420	366	357	357	380	-28.0	-24.6	5.5	-9.5
Utah	434	476	204	368	335	353	363	375	333	376	391	369	353	316	310	-28.6	-18.7	10.8	-20.7
Washington	985	928	838	748	783	713	683	678	655	625	557	604	540	627	618	-37.3	-27.6	-21.9	11.0
Wyoming	102	93	101	83	72	83	66	54	61	57	57	50	62	66	46	-54.9	-18.6	-31.3	-19.3
U.S. Service Schools	84	66	23	37	61	50	26	40	26	29	29	23	39	38	35	-58.3	-40.5	-42.0	20.7
Outlying Areas	638	588	711	730	741	830	788	901	999	960	985	926	927	870	870	36.4	30.1	18.7	-11.7
American Samoa	—	—	—	—	—	—	—	—	—	—	—	—	—	—	—	(1)	(1)	(1)	(1)
Guam	7	3	2	6	5	2	9	0	8	4	5	7	8	7	2	-71.4	-71.4	150.0	-60.0
Northern Marianas	—	—	—	—	—	—	—	—	—	—	—	—	—	—	—	(1)	(1)	(1)	(1)
Puerto Rico	626	582	707	722	730	826	777	893	988	949	980	917	913	858	862	37.7	31.9	18.6	-12.0
Trust Territories	—	—	—	—	—	—	—	—	—	—	—	—	—	—	—	(1)	(1)	(1)	(1)
Virgin Islands	5	3	2	2	6	2	2	8	3	7	0	2	6	5	6	20.0	-60.0	-100.0	(1)

[1] Insufficient data for calculating a percent change.

—Data not available.

SOURCE: U.S. Department of Education, National Center for Education Statistics, HEGIS, "Degrees and Other Formal Awards Conferred" surveys; and IPEDS, "Completions" surveys.

Table 5-12.—Bachelor's degrees conferred in physical sciences, by region and state: 1975-76 to 1989-90

Region and state	1975-76	1976-77	1977-78	1978-79	1979-80	1980-81	1981-82	1982-83	1983-84	1984-85	1985-86	1986-87	1987-88	1988-89	1989-90	Percent change: 1975-76 to 1989-90	Percent change: 1975-76 to 1980-81	Percent change: 1980-81 to 1985-86	Percent change: 1985-86 to 1989-90
United States . . .	21,465	22,497	22,986	23,207	23,410	23,952	24,052	23,405	23,671	23,732	21,731	20,070	17,806	17,186	16,131	-24.8	11.6	-9.3	-25.8
Northeast	5,942	6,155	6,339	6,312	5,977	6,236	6,187	5,845	5,940	5,988	5,205	4,945	4,437	4,275	4,009	-32.5	4.9	-16.5	-23.0
Connecticut . .	288	298	339	312	341	324	264	234	240	272	243	235	190	182	169	-41.3	12.5	-25.0	-30.5
Maine	94	94	114	90	82	118	130	146	117	102	122	105	88	93	66	-29.8	25.5	3.4	-45.9
Massachusetts	1,088	1,045	1,091	1,026	1,034	1,001	957	949	1,007	1,007	843	837	762	721	735	-32.4	-8.0	-15.8	-12.8
New Hampshire	162	203	205	178	162	182	187	161	167	153	109	112	101	100	71	-56.2	12.3	-40.1	-34.9
New Jersey . . .	473	576	557	698	506	548	545	477	459	498	477	523	345	366	314	-33.6	15.9	-13.0	-34.2
New York	2,088	2,106	2,147	2,062	1,895	1,950	2,045	1,954	2,038	2,003	1,728	1,596	1,446	1,374	1,222	-41.5	-6.6	-11.4	-29.3
Pennsylvania .	1,577	1,638	1,694	1,749	1,765	1,898	1,848	1,682	1,689	1,714	1,511	1,404	1,362	1,305	1,315	-16.6	20.4	-20.4	-13.0
Rhode Island .	75	84	104	112	88	87	79	95	107	106	73	60	67	57	47	-37.3	16.0	-16.1	-35.6
Vermont	97	111	88	85	104	128	132	147	116	133	99	73	76	77	70	-27.8	32.0	-22.7	-29.3
Midwest	5,635	5,588	5,688	5,665	5,812	5,792	5,662	5,865	6,175	6,032	5,711	5,265	4,916	4,749	4,375	-22.4	2.8	-1.4	-23.4
Illinois	908	882	863	920	997	967	921	922	1,089	985	902	818	748	740	697	-23.2	6.5	-6.7	-22.7
Indiana	640	594	606	644	676	654	627	653	681	650	678	564	502	504	430	-32.8	2.2	3.7	-36.6
Iowa	239	235	249	227	255	258	245	272	280	268	280	258	290	234	262	9.6	7.9	8.5	-6.4
Kansas	299	262	249	230	272	228	236	329	321	327	278	278	223	198	155	-48.2	-23.7	21.9	-44.2
Michigan	798	783	847	824	838	855	802	839	858	892	792	739	663	602	567	-28.9	7.1	-7.4	-28.4
Minnesota	553	580	564	569	539	586	544	611	608	558	552	505	510	511	435	-21.3	6.0	-5.8	-21.2
Missouri	470	510	491	437	438	464	485	393	436	428	398	362	345	296	303	-35.5	-1.3	-14.2	-23.9
Nebraska	144	127	155	138	150	171	169	153	159	176	139	159	140	147	124	-13.9	18.8	-18.7	-10.8
North Dakota .	54	43	57	53	48	51	58	63	97	118	117	97	79	74	74	37.0	-5.6	129.4	-36.8
Ohio	991	1,019	1,049	1,071	1,064	1,073	1,073	1,055	1,012	1,033	1,006	946	896	897	821	-17.2	8.3	-6.2	-18.4
South Dakota .	65	69	79	73	80	62	70	65	74	72	63	58	61	64	31	-52.3	-4.6	1.6	-50.8
Wisconsin	474	484	479	479	455	423	432	510	560	525	506	481	459	482	476	0.4	-10.8	19.6	-5.9
South	5,867	6,316	6,530	6,938	7,022	7,580	7,672	7,094	7,019	6,854	6,227	5,577	4,969	4,820	4,540	-22.6	29.2	-17.8	-27.1
Alabama	290	284	328	317	316	359	350	358	300	327	328	302	269	225	241	-16.9	23.8	-8.6	-26.5
Arkansas	198	205	173	188	183	200	210	207	196	171	145	141	113	125	112	-43.4	1.0	-27.5	-22.8
Delaware	60	84	93	83	77	77	81	108	134	120	110	99	80	68	68	13.3	28.3	42.9	-38.2
District of Columbia . . .	153	146	111	150	137	148	146	142	140	141	105	98	82	102	100	-34.6	-3.3	-29.1	-4.8
Florida	670	909	974	1,151	1,295	1,408	1,477	655	590	592	535	440	408	410	383	-42.8	110.1	-62.0	-28.4
Georgia	434	440	412	436	403	465	481	501	446	437	375	324	358	325	323	-25.6	7.1	-19.4	-13.9
Kentucky	231	220	207	211	226	248	273	279	268	311	315	246	237	214	197	-14.7	7.4	27.0	-37.5
Louisiana	237	251	305	295	279	286	349	405	417	358	406	286	246	218	174	-26.6	20.7	42.0	-57.1
Maryland	269	255	272	304	293	299	279	240	218	240	237	204	221	229	203	-24.5	11.2	-20.7	-14.3
Mississippi . . .	185	181	164	205	200	206	203	212	223	241	183	167	140	169	150	-18.9	11.4	-11.2	-18.0
North Carolina	648	713	741	750	718	750	687	732	769	716	691	570	604	545	520	-19.8	15.7	-7.9	-24.7
Oklahoma	268	290	287	302	304	366	333	346	351	321	278	281	248	218	235	-12.3	36.6	-24.0	-15.5
South Carolina	235	217	225	269	242	256	222	261	212	239	235	199	166	199	192	-18.3	8.9	-8.2	-18.3
Tennessee . . .	401	439	393	418	387	406	385	382	407	379	331	348	288	266	251	-37.4	1.2	-18.5	-24.2
Texas	918	1,001	1,084	1,110	1,169	1,308	1,334	1,413	1,504	1,441	1,198	1,106	877	828	777	-15.4	42.5	-8.4	-35.1
Virginia	530	518	562	588	623	625	669	660	637	635	570	553	495	521	511	-3.6	17.9	-8.8	-10.4
West Virginia .	140	163	199	161	170	173	193	193	207	185	185	213	137	158	103	-26.4	23.6	6.9	-44.3
West	3,756	4,137	4,109	3,952	4,189	4,002	4,209	4,324	4,249	4,411	4,230	3,739	3,199	3,087	2,982	-20.6	6.5	5.7	-29.5
Alaska	15	6	9	15	18	15	11	24	13	27	26	28	18	24	23	53.3	0.0	73.3	-11.5
Arizona	195	214	235	237	232	213	256	238	238	255	222	186	185	169	165	-15.4	9.2	4.2	-25.7
California	1,859	1,854	1,887	1,849	2,074	1,843	1,898	2,018	1,940	2,088	2,021	1,859	1,598	1,515	1,517	-18.4	-0.9	9.7	-24.9
Colorado	402	431	477	456	461	496	542	557	608	550	534	376	313	328	298	-25.9	23.4	7.7	-44.2
Hawaii	42	47	47	42	47	54	35	45	44	43	40	53	36	30	35	-16.7	28.6	-25.9	-12.5
Idaho	64	95	76	73	87	71	70	117	98	79	69	81	65	39	61	-4.7	10.9	-2.8	-11.6

Table 5-12.—Continued

Region and state	1975-76	1976-77	1977-78	1978-79	1979-80	1980-81	1981-82	1982-83	1983-84	1984-85	1985-86	1986-87	1987-88	1988-89	1989-90	Percent change: 1975-76 to 1989-90	Percent change: 1975-76 to 1980-81	Percent change: 1980-81 to 1985-86	Percent change: 1985-86 to 1989-90
West Continued																			
Montana	80	116	107	125	113	128	128	139	124	136	125	96	75	69	47	-41.3	60.0	-2.3	-62.4
Nevada	31	47	55	49	39	53	50	53	50	47	55	68	28	37	29	-6.5	71.0	3.8	-47.3
New Mexico	136	146	140	138	124	144	143	125	132	160	141	111	115	104	86	-36.8	5.9	-2.1	-39.0
Oregon	237	401	332	240	243	243	251	273	291	285	288	258	190	173	186	-21.5	2.5	18.5	-35.4
Utah	219	276	270	226	245	250	263	244	243	242	258	232	189	208	186	-15.1	14.2	3.2	-27.9
Washington	440	465	424	469	451	450	503	436	417	436	409	349	352	370	333	-24.3	2.3	-9.1	-18.6
Wyoming	36	39	50	33	55	42	59	55	51	63	42	42	35	21	16	-55.6	16.7	0.0	-61.9
U.S. Service Schools	265	301	320	340	410	342	322	277	288	447	358	544	285	255	225	-15.1	29.1	4.7	-37.2
Outlying Areas	94	121	189	156	251	223	320	263	251	250	252	190	156	199	198	110.6	137.2	13.0	-21.4
American Samoa	—	—	—	—	—	—	—	—	—	—	—	—	—	—	—	(1)	(1)	(1)	(1)
Guam	3	2	10	—	—	4	1	2	1	—	1	—	2	1	—	(1)	33.3	-75.0	(1)
Northern Marianas	—	—	—	—	—	—	—	—	—	—	—	—	—	—	—	(1)	(1)	(1)	(1)
Puerto Rico	91	115	178	153	250	214	317	259	246	245	247	189	153	195	197	116.5	135.2	15.4	-20.2
Trust Territories	—	—	—	—	—	—	—	—	—	—	—	—	—	—	—	(1)	(1)	(1)	(1)
Virgin Islands	—	4	1	3	1	5	2	2	4	5	4	1	1	3	1	(1)	(1)	-20.0	-75.0

[1] Insufficient data for calculating a percent change.

—Data not available.

SOURCE: U.S. Department of Education, National Center for Education Statistics, HEGIS, "Degrees and Other Formal Awards Conferred" surveys; and IPEDS, "Completions" surveys.

Table 5-13.—Master's degrees conferred in agricultural sciences, by region and state: 1975-76 to 1989-90

Region and state	1975-76	1976-77	1977-78	1978-79	1979-80	1980-81	1981-82	1982-83	1983-84	1984-85	1985-86	1986-87	1987-88	1988-89	1989-90	Percent change 1975-76 to 1989-90	Percent change 1975-76 to 1980-81	Percent change 1980-81 to 1985-86	Percent change 1985-86 to 1989-90
United States	3,340	3,724	4,023	3,994	3,976	4,003	4,163	4,254	4,178	3,928	3,801	3,522	3,479	3,245	3,373	1.0	19.9	-5.0	-11.3
Northeast	431	384	430	441	438	427	463	436	409	424	423	385	424	395	435	0.9	-0.9	-0.9	2.8
Connecticut	72	72	73	64	77	63	92	89	74	100	107	73	72	70	72	0.0	-12.5	69.8	-32.7
Maine	24	14	19	28	28	24	21	25	14	36	29	34	16	15	30	25.0	0.0	20.8	3.4
Massachusetts	66	50	59	59	47	47	46	39	45	27	26	27	37	32	33	-50.0	-28.8	-44.7	26.9
New Hampshire	17	23	15	25	14	17	27	28	28	24	15	22	17	28	40	135.3	0.0	-11.8	166.7
New Jersey	45	38	46	41	46	43	46	23	40	32	36	31	43	44	44	-2.2	-4.4	-16.3	22.2
New York	125	112	111	96	122	129	142	131	122	112	126	123	154	129	138	10.4	3.2	-2.3	9.5
Pennsylvania	56	54	62	89	54	79	61	71	63	63	54	50	48	47	44	-21.4	41.1	-31.6	-18.5
Rhode Island	13	11	19	19	25	12	14	6	9	11	11	9	15	14	15	15.4	-7.7	-8.3	36.4
Vermont	13	10	26	20	25	13	14	24	14	19	19	16	22	16	19	46.2	0.0	46.2	0.0
Midwest	950	1,101	1,245	1,148	1,121	1,119	1,149	1,234	1,193	1,073	1,100	946	960	886	1,017	7.1	17.8	-1.7	-7.5
Illinois	137	158	140	126	142	159	147	149	158	137	163	146	118	124	141	2.9	16.1	2.5	-13.5
Indiana	58	87	85	82	80	87	84	93	79	74	74	59	62	34	59	1.7	50.0	-14.9	-20.3
Iowa	64	77	94	98	70	94	80	90	103	89	85	65	47	57	67	4.7	46.9	-9.6	-21.2
Kansas	41	58	112	75	79	59	79	65	64	52	69	50	53	55	63	53.7	43.9	16.9	-8.7
Michigan	221	220	224	197	220	195	209	215	171	158	172	177	167	159	194	-12.2	-11.8	-11.8	12.8
Minnesota	66	86	87	93	86	95	102	108	102	94	90	84	94	95	104	57.6	43.9	-5.3	15.6
Missouri	61	84	88	85	101	91	79	106	102	83	70	72	87	53	66	8.2	49.2	-23.1	-5.7
Nebraska	40	44	68	59	52	73	54	61	84	61	55	54	65	53	62	55.0	82.5	-24.7	12.7
North Dakota	26	25	39	27	25	32	43	42	41	38	26	41	36	33	33	26.9	23.1	-18.8	26.9
Ohio	93	98	96	139	92	105	98	127	125	106	90	69	87	87	86	-7.5	12.9	-14.3	-4.4
South Dakota	19	20	20	18	27	24	24	18	28	31	38	19	19	26	32	68.4	26.3	58.3	-15.8
Wisconsin	124	144	192	149	147	105	150	160	136	150	168	110	125	110	110	-11.3	-15.3	60.0	-34.5
South	1,066	1,243	1,288	1,357	1,391	1,381	1,389	1,441	1,533	1,373	1,281	1,267	1,138	1,116	1,042	-2.3	29.5	-7.2	-18.7
Alabama	57	66	62	97	101	76	114	118	127	103	90	95	91	96	95	66.7	33.3	18.4	5.6
Arkansas	29	48	39	54	55	66	67	66	76	72	65	57	64	60	58	100.0	127.6	-1.5	-10.8
Delaware	11	8	14	17	10	16	20	20	24	17	14	28	16	16	16	45.5	45.5	-12.5	14.3
District of Columbia	—	21	19	22	0	0	0	0	0	0	0	—	—	—	—	(1)	(1)	(1)	(1)
Florida	88	101	101	116	98	98	113	111	110	106	96	96	84	88	75	-14.8	11.4	-2.0	-21.9
Georgia	50	59	65	67	72	81	74	85	95	97	81	63	72	74	65	30.0	62.0	0.0	-19.8
Kentucky	52	40	63	53	81	71	62	54	79	73	80	80	79	64	52	0.0	36.5	12.7	-35.0
Louisiana	62	56	75	75	68	65	50	66	68	75	54	54	45	34	38	-38.7	4.8	-16.9	-29.6
Maryland	48	39	55	51	44	65	44	61	47	48	50	45	25	30	32	-33.3	35.4	-23.1	-36.0
Mississippi	66	69	78	66	80	80	51	76	65	64	87	71	67	79	81	22.7	21.2	8.8	-6.9
North Carolina	95	109	120	101	122	112	132	126	140	132	129	140	130	110	100	5.3	17.9	15.2	-22.5
Oklahoma	51	52	46	79	65	57	67	48	48	57	52	39	50	42	41	-19.6	11.8	-8.8	-21.2
South Carolina	20	42	28	40	34	35	44	44	44	43	43	27	38	27	21	5.0	75.0	22.9	-51.2
Tennessee	99	108	86	96	102	95	103	80	74	78	67	69	59	74	51	-48.5	-4.0	-29.5	-23.9
Texas	248	314	331	323	348	338	318	346	381	310	299	289	237	229	251	1.2	36.3	-11.5	-16.1
Virginia	54	73	67	65	81	82	86	87	114	61	40	87	59	64	56	3.7	51.9	-51.2	40.0
West Virginia	36	38	39	35	30	44	44	53	41	37	34	27	22	29	10	-72.2	22.2	-22.7	-70.6
West	893	996	1,060	1,048	1,026	1,076	1,162	1,143	1,043	1,058	997	924	957	848	879	-1.6	20.5	-7.3	-11.8
Alaska	5	10	7	8	10	13	16	12	14	25	20	17	20	14	17	240.0	160.0	53.8	-15.0
Arizona	55	98	74	84	64	80	72	78	89	91	93	75	69	85	82	49.1	45.5	16.3	-11.8
California	293	304	301	327	308	310	378	384	324	314	295	270	278	231	253	-13.7	5.8	-4.8	-14.2
Colorado	78	85	109	97	105	101	105	114	113	105	98	78	101	71	71	-9.0	29.5	-3.0	-27.6
Hawaii	33	24	21	28	28	35	36	37	35	36	23	17	25	13	19	-42.4	6.1	-34.3	-17.4
Idaho	37	56	66	47	53	68	49	60	56	57	63	58	41	50	51	37.8	83.8	-7.4	-19.0

Table 5-13.—Continued

Region and state	1975-76	1976-77	1977-78	1978-79	1979-80	1980-81	1981-82	1982-83	1983-84	1984-85	1985-86	1986-87	1987-88	1988-89	1989-90	Percent change 1975-76 to 1989-90	1975-76 to 1980-81	1980-81 to 1985-86	1985-86 to 1989-90
West Continued																			
Montana	48	38	46	42	38	46	46	51	52	55	52	50	49	49	37	-22.9	-4.2	13.0	-28.8
Nevada	15	24	17	24	12	22	12	11	6	10	26	13	16	15	19	26.7	46.7	18.2	-26.9
New Mexico	60	53	56	59	60	73	72	45	52	71	54	50	66	55	70	16.7	21.7	-26.0	29.6
Oregon	74	86	108	94	92	131	108	104	103	111	85	107	103	110	101	36.5	77.0	-35.1	18.8
Utah	84	75	97	64	97	75	98	70	64	64	52	52	58	51	47	-44.0	-10.7	-30.7	-9.6
Washington	98	116	139	151	136	97	151	158	108	104	114	118	111	84	94	-4.1	-1.0	17.5	-17.5
Wyoming	13	27	19	23	23	25	19	19	27	15	22	19	20	20	18	38.5	92.3	-12.0	-18.2
U.S. Service Schools	—	—	—	—	—	—	—	—	—	—	—	—	—	—	—	(1)	(1)	(1)	(1)
Outlying Areas	11	11	13	13	11	14	15	18	24	22	38	49	18	25	18	63.6	27.3	171.4	-52.6
American Samoa	—	—	—	—	—	—	—	—	—	—	—	—	—	—	—	(1)	(1)	(1)	(1)
Guam	—	—	—	0	—	0	0	0	0	0	0	—	0	0	—	(1)	(1)	(1)	(1)
Northern Marianas	—	—	—	—	—	—	—	—	—	—	—	—	—	—	—	(1)	(1)	(1)	(1)
Puerto Rico	11	11	13	13	11	14	15	18	24	22	38	49	18	25	18	63.6	27.3	171.4	-52.6
Trust Territories	—	—	—	—	—	—	—	0	0	0	0	—	—	—	—	(1)	(1)	(1)	(1)
Virgin Islands	—	—	—	—	—	—	—	0	0	0	0	—	—	—	—	(1)	(1)	(1)	(1)

[1] Insufficient data for calculating a percent change.

—Data not available.

SOURCE: U.S. Department of Education, National Center for Education Statistics, HEGIS, "Degrees and Other Formal Awards Conferred" surveys; and IPEDS, "Completions" surveys.

Table 5-14.—Master's degrees conferred in computer sciences, by region and state: 1975-76 to 1989-90

Region and state	1975-76	1976-77	1977-78	1978-79	1979-80	1980-81	1981-82	1982-83	1983-84	1984-85	1985-86	1986-87	1987-88	1988-89	1989-90	Percent change 1975-76 to 1989-90	1975-76 to 1980-81	1980-81 to 1985-86	1985-86 to 1989-90
United States . . .	2,603	2,798	3,038	3,055	3,647	4,218	4,935	5,321	6,190	7,101	8,070	8,481	9,197	9,414	9,643	270.5	62.0	91.3	19.5
Northeast	752	852	934	889	1,042	1,279	1,651	1,941	2,252	2,606	3,074	3,314	3,246	3,439	3,513	367.2	70.1	140.3	14.3
Connecticut . .	91	94	92	80	74	95	110	129	129	168	208	216	196	196	205	125.3	4.4	118.9	-1.4
Maine	—	0	0	0	0	0	0	0	0	1	0	1	2	3	4	([1])	([1])	([1])	([1])
Massachusetts	62	60	69	86	79	127	162	192	201	217	291	343	362	384	372	500.0	104.8	129.1	27.8
New Hampshire	0	0	0	0	1	1	17	38	30	35	40	37	40	54	66	([1])	([1])	3900.0	65.0
New Jersey . . .	146	131	154	149	187	260	347	368	364	366	492	533	567	682	714	389.0	78.1	89.2	45.1
New York.	302	420	440	394	433	549	652	846	1,034	1,360	1,457	1,586	1,535	1,551	1,602	430.5	81.8	165.4	10.0
Pennsylvania .	144	127	166	158	248	220	339	344	450	414	533	532	481	518	493	242.4	52.8	142.3	-7.5
Rhode Island .	7	15	6	16	5	17	20	9	30	37	46	56	57	46	53	657.1	142.9	170.6	15.2
Vermont	—	5	7	6	15	10	4	15	14	8	7	10	6	5	4	([1])	([1])	-30.0	-42.9
Midwest	598	625	662	732	798	1,001	1,111	1,212	1,334	1,335	1,470	1,557	1,639	1,736	1,783	198.2	67.4	46.9	21.3
Illinois	133	148	216	180	212	267	352	352	392	428	488	588	660	691	741	457.1	100.8	82.8	51.8
Indiana	79	70	77	73	108	99	123	142	179	132	118	122	103	118	104	31.6	25.3	19.2	-11.9
Iowa.	33	26	27	25	31	20	24	31	38	45	51	71	62	47	57	72.7	-39.4	155.0	11.8
Kansas	26	29	45	33	26	36	70	53	60	78	81	75	67	55	43	65.4	38.5	125.0	-46.9
Michigan	105	106	84	162	153	208	134	173	196	218	199	206	231	241	229	118.1	98.1	-4.3	15.1
Minnesota	16	16	19	14	14	23	18	20	38	33	50	69	70	105	103	543.8	43.8	117.4	106.0
Missouri	23	39	48	53	52	53	59	109	125	118	144	115	107	102	153	565.2	130.4	171.7	6.3
Nebraska	14	10	14	14	9	15	23	17	21	11	24	20	37	39	21	50.0	7.1	60.0	-12.5
North Dakota .	2	0	1	1	1	12	6	9	8	23	19	18	19	19	19	850.0	500.0	58.3	0.0
Ohio.	100	114	92	114	116	188	162	167	171	182	188	176	192	211	205	105.0	88.0	0.0	9.0
South Dakota .	0	0	0	0	0	0	0	0	1	3	1	6	4	7	5	([1])	([1])	([1])	400.0
Wisconsin	67	67	39	63	76	80	140	139	105	64	107	91	87	101	103	53.7	19.4	33.8	-3.7
South	711	752	836	819	1,016	1,070	1,125	1,308	1,654	2,037	2,242	2,370	2,714	2,561	2,570	261.5	50.5	109.5	14.6
Alabama	7	11	17	20	22	31	45	45	71	109	81	86	108	95	118	1585.7	342.9	161.3	45.7
Arkansas.	1	3	1	0	4	5	12	19	21	16	3	10	18	9	13	1200.0	400.0	-40.0	333.3
Delaware.	7	2	4	5	5	4	7	7	7	10	21	22	20	34	34	385.7	-42.9	425.0	61.9
District of Columbia . . .	246	271	304	221	248	191	167	217	187	217	234	209	215	214	215	-12.6	-22.4	22.5	-8.1
Florida	23	29	36	63	80	86	141	195	224	246	249	268	329	320	295	1182.6	273.9	189.5	18.5
Georgia	73	82	82	84	75	102	104	94	110	142	145	176	212	186	133	82.2	39.7	42.2	-8.3
Kentucky	18	13	21	23	17	16	18	23	15	25	23	32	39	44	48	166.7	-11.1	43.8	108.7
Louisiana	14	18	20	15	16	25	26	23	37	59	128	114	109	95	86	514.3	78.6	412.0	-32.8
Maryland	66	64	80	96	113	128	147	150	245	279	297	321	397	442	503	662.1	93.9	132.0	69.4
Mississippi . . .	10	21	19	17	49	56	48	51	73	104	121	137	135	93	68	580.0	460.0	116.1	-43.8
North Carolina	13	13	15	15	30	37	40	42	48	62	76	63	105	63	77	492.3	184.6	105.4	1.3
Oklahoma	32	22	16	26	38	28	23	18	18	40	33	31	41	51	44	37.5	-12.5	17.9	33.3
South Carolina	0	2	8	6	11	15	14	29	26	31	34	46	57	46	52	([1])	([1])	126.7	52.9
Tennessee . . .	18	27	29	22	46	41	39	36	47	51	56	55	65	60	66	266.7	127.8	36.6	17.9
Texas.	136	136	147	167	221	241	218	266	396	493	543	566	596	506	482	254.4	77.2	125.3	-11.2
Virginia	44	32	31	32	31	52	62	65	103	135	172	194	236	272	307	597.7	18.2	230.8	78.5
West Virginia .	3	6	6	7	10	12	14	28	26	18	26	40	32	31	29	866.7	300.0	116.7	11.5
West	484	495	537	540	689	793	933	792	895	1,081	1,232	1,211	1,587	1,649	1,678	246.7	63.8	55.4	36.2
Alaska	0	—	0	0	—	—	0	0	0	0	0	2	3	3	2	([1])	([1])	([1])	([1])
Arizona	27	28	37	34	27	46	57	19	39	59	65	80	76	90	86	218.5	70.4	41.3	32.3
California	355	328	363	359	502	562	669	583	646	673	804	795	1,020	1,100	1,113	213.5	58.3	43.1	38.4
Colorado	7	21	26	30	36	48	52	45	47	62	97	93	148	157	178	2442.9	585.7	102.1	83.5
Hawaii	20	19	15	18	13	17	14	14	11	28	18	21	12	12	17	-15.0	-15.0	5.9	-5.6
Idaho	2	1	0	0	2	0	5	9	10	12	12	14	6	1	4	100.0	-100.0	([1])	-66.7

Table 5-14.—Continued

Region and state	1975-76	1976-77	1977-78	1978-79	1979-80	1980-81	1981-82	1982-83	1983-84	1984-85	1985-86	1986-87	1987-88	1988-89	1989-90	Percent change 1975-76 to 1989-90	Percent change 1975-76 to 1980-81	Percent change 1980-81 to 1985-86	Percent change 1985-86 to 1989-90
West Continued																			
Montana	0	0	0	0	0	0	1	4	7	14	9	9	21	12	4	(1)	(1)	(1)	-55.6
Nevada	—	—	—	—	0	0	0	0	0	1	1	6	2	3	11	(1)	(1)	(1)	1000.0
New Mexico ..	14	19	22	15	23	22	21	12	31	51	37	30	37	42	27	92.9	57.1	68.2	-27.0
Oregon	22	25	32	40	35	23	34	29	31	52	36	31	53	50	55	150.0	4.5	56.5	52.8
Utah.........	11	22	15	20	20	27	17	14	20	40	62	63	83	66	68	518.2	145.5	129.6	9.7
Washington...	26	32	27	24	31	43	61	58	46	82	88	61	113	108	104	300.0	65.4	104.7	18.2
Wyoming.....	0	0	0	0	0	5	2	5	7	7	3	6	13	5	9	(1)	(1)	-40.0	200.0
U.S. Service Schools......	58	74	69	75	102	75	115	68	55	42	52	29	11	29	99	70.7	29.3	-30.7	90.4
Outlying Areas..	0	0	0	0	0	0	0	0	0	0	0	0	0	7	0	(1)	(1)	(1)	(1)
American Samoa.....	—	—	—	—	—	—	—	—	—	—	—	—	—	—	—	(1)	(1)	(1)	(1)
Guam	—	—	—	—	—	—	—	0	0	0	0	—	—	—	—	(1)	(1)	(1)	(1)
Northern Marianas...	—	—	—	—	—	—	—	—	—	—	—	—	—	—	—	(1)	(1)	(1)	(1)
Puerto Rico ..	0	0	0	0	0	0	0	0	0	0	0	—	—	7	0	(1)	(1)	(1)	(1)
Trust Territories	—	—	—	—	—	—	—	—	—	—	—	—	—	—	—	(1)	(1)	(1)	(1)
Virgin Islands .	—	—	—	—	—	—	—	—	—	0	—	—	—	—	—	(1)	(1)	(1)	(1)

[1] Insufficient data for calculating a percent change.

—Data not available.

SOURCE: U.S. Department of Education, National Center for Education Statistics, HEGIS, "Degrees and Other Formal Awards Conferred" surveys; and IPEDS, "Completions" surveys.

Table 5-15.—Master's degrees conferred in engineering, by region and state: 1975-76 to 1989-90

Region and state	1975-76	1976-77	1977-78	1978-79	1979-80	1980-81	1981-82	1982-83	1983-84	1984-85	1985-86	1986-87	1987-88	1988-89	1989-90	Percent change 1975-76 to 1989-90	Percent change 1975-76 to 1980-81	Percent change 1980-81 to 1985-86	Percent change 1985-86 to 1989-90
United States . . .	16,342	16,245	16,398	15,495	16,243	16,709	17,939	19,350	20,661	21,557	21,661	22,658	23,388	24,572	24,848	52.0	2.2	29.6	14.7
Northeast	4,653	4,525	4,699	4,333	4,502	4,604	5,075	5,417	5,692	5,722	5,606	6,104	6,063	6,604	6,449	38.6	-1.1	21.8	15.0
Connecticut . .	172	180	171	156	181	200	240	234	270	266	329	282	282	299	314	82.6	16.3	64.5	-4.6
Maine	20	30	20	23	15	12	11	10	18	14	14	18	13	16	29	45.0	-40.0	16.7	107.1
Massachusetts	1,186	1,183	1,210	1,071	1,189	1,229	1,280	1,378	1,375	1,428	1,444	1,603	1,613	1,710	1,685	42.1	3.6	17.5	16.7
New Hampshire	33	26	31	20	37	47	46	57	60	68	55	71	66	52	88	166.7	42.4	17.0	60.0
New Jersey . . .	559	482	505	464	481	443	496	518	641	592	623	684	819	954	917	64.0	-20.8	40.6	47.2
New York	1,752	1,707	1,831	1,678	1,673	1,716	1,941	2,032	2,096	2,100	1,923	2,186	1,993	2,238	2,180	24.4	-2.1	12.1	13.4
Pennsylvania .	822	841	838	827	827	874	956	1,071	1,098	1,122	1,090	1,143	1,159	1,215	1,150	39.9	6.3	24.7	5.5
Rhode Island .	58	48	57	55	63	58	67	72	75	80	80	83	83	83	72	24.1	0.0	37.9	-10.0
Vermont	51	28	36	39	36	25	38	45	59	52	48	34	35	37	14	-72.5	-51.0	92.0	-70.8
Midwest	3,787	3,873	3,922	3,693	3,947	3,871	4,369	4,563	4,921	5,226	5,068	5,091	5,242	5,495	5,662	49.5	2.2	30.9	11.7
Illinois	797	734	767	721	817	737	848	820	932	942	872	933	888	992	993	24.6	-7.5	18.3	13.9
Indiana	356	421	372	333	337	364	391	429	467	503	472	483	499	508	487	36.8	2.2	29.7	3.2
Iowa	148	181	169	142	194	179	201	166	225	245	193	196	227	207	251	69.6	20.9	7.8	30.1
Kansas	186	184	205	184	187	173	171	184	193	212	211	222	223	252	274	47.3	-7.0	22.0	29.9
Michigan	735	710	795	742	768	784	971	945	1,047	1,090	1,136	1,028	1,109	1,180	1,134	54.3	6.7	44.9	-0.2
Minnesota	100	120	136	119	106	116	146	157	137	176	193	157	207	211	245	145.0	16.0	66.4	26.9
Missouri	419	415	418	357	356	347	392	494	474	567	487	470	501	488	597	42.5	-17.2	40.3	22.6
Nebraska	53	37	67	49	53	54	59	78	78	86	74	58	53	64	62	17.0	1.9	37.0	-16.2
North Dakota .	25	27	15	26	19	21	26	35	50	44	43	37	44	39	45	80.0	-16.0	104.8	4.7
Ohio	658	668	618	599	719	706	757	807	820	855	941	1,012	1,006	1,030	1,049	59.4	7.3	33.3	11.5
South Dakota .	37	51	56	57	47	59	53	63	67	87	90	64	76	77	70	89.2	59.5	52.5	-22.2
Wisconsin	273	325	304	364	344	331	354	385	431	419	356	431	409	447	455	66.7	21.2	7.6	27.8
South	3,887	3,779	3,749	3,670	3,836	4,160	4,166	4,683	4,977	5,337	5,730	5,860	6,185	6,384	6,580	69.3	7.0	37.7	14.8
Alabama	102	124	109	120	133	89	126	131	152	150	168	223	241	248	327	220.6	-12.7	88.8	94.6
Arkansas	197	157	158	149	136	158	199	200	216	209	261	221	190	57	69	-65.0	-19.8	65.2	-73.6
Delaware	48	28	36	43	32	41	51	47	48	55	57	45	56	50	50	4.2	-14.6	39.0	-12.3
District of Columbia . . .	301	286	271	240	353	350	304	510	306	406	321	377	422	472	474	57.5	16.3	-8.3	47.7
Florida	361	326	317	269	297	285	319	369	390	445	515	557	627	678	748	107.2	-21.1	80.7	45.2
Georgia	297	269	280	292	312	381	354	340	421	435	409	455	485	463	506	70.4	28.3	7.3	23.7
Kentucky	144	165	145	147	128	178	161	181	208	160	215	220	236	275	227	57.6	23.6	20.8	5.6
Louisiana	110	144	144	125	127	148	183	192	250	321	282	267	289	225	220	100.0	34.5	90.5	-22.0
Maryland	194	156	152	153	180	181	198	212	239	275	341	374	395	468	550	183.5	-6.7	88.4	61.3
Mississippi . . .	41	45	51	71	58	81	76	67	77	122	94	114	108	114	95	131.7	97.6	16.0	1.1
North Carolina	168	191	143	154	135	189	183	169	250	250	283	314	332	300	312	85.7	12.5	49.7	10.2
Oklahoma	289	272	239	259	209	251	191	223	223	297	324	305	320	336	292	1.0	-13.1	29.1	-9.9
South Carolina	111	124	131	116	133	130	123	140	167	200	188	194	183	194	214	92.8	17.1	44.6	13.8
Tennessee . . .	250	279	268	232	248	269	238	251	233	248	306	275	309	338	311	24.4	7.6	13.8	1.6
Texas	890	799	854	870	931	997	966	1,094	1,239	1,239	1,396	1,300	1,312	1,442	1,461	64.2	12.0	40.0	4.7
Virginia	281	346	330	339	320	333	371	392	409	388	430	501	555	602	608	116.4	18.5	29.1	41.4
West Virginia .	103	68	121	91	104	99	123	165	149	137	140	118	125	122	116	12.6	-3.9	41.4	-17.1
West	3,684	3,703	3,641	3,543	3,655	3,723	3,977	4,319	4,638	4,840	4,778	5,246	5,513	5,677	5,946	61.4	1.1	28.3	24.4
Alaska	13	22	11	14	20	31	16	32	26	37	38	42	51	41	46	253.8	138.5	22.6	21.1
Arizona	202	192	183	176	167	172	283	291	272	325	321	337	387	389	428	111.9	-14.9	86.6	33.3
California	2,453	2,474	2,400	2,333	2,421	2,395	2,555	2,792	2,989	3,061	3,046	3,324	3,425	3,508	3,661	49.2	-2.4	27.2	20.2
Colorado	205	208	267	206	223	249	288	284	326	287	287	394	421	444	507	147.3	21.5	15.3	76.7
Hawaii	53	50	31	39	35	37	30	32	27	37	21	43	50	42	30	-43.4	-30.2	-43.2	42.9
Idaho	65	37	50	56	44	49	45	58	43	56	57	53	50	64	85	30.8	-24.6	16.3	49.1

Table 5-15.—Continued

Region and state	1975-76	1976-77	1977-78	1978-79	1979-80	1980-81	1981-82	1982-83	1983-84	1984-85	1985-86	1986-87	1987-88	1988-89	1989-90	Percent change 1975-76 to 1989-90	Percent change 1975-76 to 1980-81	Percent change 1980-81 to 1985-86	Percent change 1985-86 to 1989-90
West Continued																			
Montana	44	62	28	41	32	43	49	58	59	62	63	63	67	70	66	50.0	-2.3	46.5	4.8
Nevada	16	21	5	9	16	18	11	10	21	24	36	32	44	58	46	187.5	12.5	100.0	27.8
New Mexico ..	104	94	110	119	128	141	121	137	142	125	141	134	132	197	187	79.8	35.6	0.0	32.6
Oregon	96	112	115	100	119	113	155	121	157	140	151	151	164	152	165	71.9	17.7	33.6	9.3
Utah.........	180	197	202	183	189	198	170	221	287	353	254	300	355	329	273	51.7	10.0	28.3	7.5
Washington...	222	207	197	240	242	246	234	248	253	291	318	335	328	347	425	91.4	10.8	29.3	33.6
Wyoming.....	31	27	42	27	19	31	20	35	36	42	45	38	39	36	27	-12.9	0.0	45.2	-40.0
U.S. Service Schools......	331	365	387	256	303	351	352	368	433	432	479	357	385	412	211	-36.3	6.0	36.5	-55.9
Outlying Areas..	7	6	11	15	7	7	8	12	9	9	10	11	11	22	32	357.1	0.0	42.9	220.0
American Samoa.....	—	—	—	—	—	—	—	—	—	—	—	—	—	—	—	(1)	(1)	(1)	(1)
Guam	—	0	—	0	—	0	0	0	0	0	0	—	—	—	—	(1)	(1)	(1)	(1)
Northern Marianas...	—	—	—	—	—	—	—	—	—	—	—	—	—	—	—	(1)	(1)	(1)	(1)
Puerto Rico ..	7	6	11	15	7	7	8	12	9	9	10	11	11	22	32	357.1	0.0	42.9	220.0
Trust Territories	—	—	—	—	—	—	—	0	0	0	0	—	—	—	—	(1)	(1)	(1)	(1)
Virgin Islands .	—	0	0	—	—	—	—	—	—	—	—	—	—	—	—	(1)	(1)	(1)	(1)

[1] Insufficient data for calculating a percent change.

—Data not available.

SOURCE: U.S. Department of Education, National Center for Education Statistics, HEGIS, "Degrees and Other Formal Awards Conferred" surveys; and IPEDS, "Completions" surveys.

Table 5-16.—Master's degrees conferred in health sciences, by region and state: 1975-76 to 1989-90

Region and state	1975-76	1976-77	1977-78	1978-79	1979-80	1980-81	1981-82	1982-83	1983-84	1984-85	1985-86	1986-87	1987-88	1988-89	1989-90	Percent change			
																1975-76 to 1989-90	1975-76 to 1980-81	1980-81 to 1985-86	1985-86 to 1989-90
United States	11,885	12,323	13,619	14,781	15,068	16,004	15,942	17,068	17,443	17,383	18,624	18,420	18,665	19,293	20,354	71.3	34.7	16.4	9.3
Northeast	3,159	3,017	3,608	4,135	4,302	4,245	4,335	4,724	4,817	4,812	5,041	5,367	5,131	5,543	5,571	76.4	34.4	18.8	10.5
Connecticut	169	221	147	158	145	306	293	289	319	309	343	338	354	341	332	96.4	81.1	12.1	-3.2
Maine	0	0	0	0	0	2	1	2	2	6	8	15	17	29	16	([1])	([1])	300.0	100.0
Massachusetts	819	751	824	1,036	1,100	951	978	1,100	1,009	1,084	1,233	1,325	1,312	1,376	1,365	66.7	16.1	29.7	10.7
New Hampshire	0	0	0	0	0	7	2	9	11	7	18	28	23	20	27	([1])	([1])	157.1	50.0
New Jersey	176	142	120	202	215	217	204	173	197	203	223	245	226	269	261	48.3	23.3	2.8	17.0
New York	1,441	1,425	1,949	2,070	2,133	2,039	2,022	2,040	2,141	2,021	2,018	2,061	1,810	2,058	2,079	44.3	41.5	-1.0	3.0
Pennsylvania	490	425	517	617	655	649	749	953	981	1,040	1,085	1,231	1,244	1,302	1,348	175.1	32.4	67.2	24.2
Rhode Island	46	34	28	23	22	33	28	111	116	102	77	89	106	98	101	119.6	-28.3	133.3	31.2
Vermont	18	19	23	29	32	41	58	47	41	40	36	35	39	50	42	133.3	127.8	-12.2	16.7
Midwest	3,237	3,396	3,697	3,929	4,107	4,327	4,199	4,353	4,392	4,502	4,968	4,652	4,780	4,851	5,087	57.2	33.7	14.8	2.4
Illinois	637	645	756	803	803	925	941	968	1,033	1,124	1,426	1,095	1,054	1,130	1,126	76.8	45.2	54.2	-21.0
Indiana	276	270	259	357	359	373	336	406	378	370	361	354	362	394	405	46.7	35.1	-3.2	12.2
Iowa	191	177	149	145	148	163	144	131	157	136	159	137	197	176	229	19.9	-14.7	-2.5	44.0
Kansas	94	108	98	88	83	123	134	144	162	176	181	200	181	208	204	117.0	30.9	47.2	12.7
Michigan	787	693	685	752	850	874	698	684	605	634	696	655	699	695	737	-6.4	11.1	-20.4	5.9
Minnesota	308	312	371	309	351	311	311	390	393	421	387	451	394	397	421	36.7	1.0	24.4	8.8
Missouri	263	412	480	456	471	506	531	514	490	500	467	434	451	476	503	91.3	92.4	-7.7	7.7
Nebraska	35	46	50	74	84	73	67	65	40	61	60	71	51	52	40	14.3	108.6	-17.8	-33.3
North Dakota	30	31	22	29	32	25	35	39	50	46	71	75	70	59	77	156.7	-16.7	184.0	8.5
Ohio	398	465	512	568	595	583	618	663	692	656	728	753	877	844	909	128.4	46.5	24.9	24.9
South Dakota	24	19	17	32	4	18	22	40	42	24	28	29	35	17	37	54.2	-25.0	55.6	32.1
Wisconsin	194	218	298	316	327	353	362	309	350	354	404	398	409	403	399	105.7	82.0	14.4	-1.2
South	3,209	3,542	3,862	4,062	4,191	4,458	4,481	4,703	4,947	4,917	5,459	5,265	5,598	5,754	5,817	81.3	38.9	22.5	6.6
Alabama	124	136	186	235	295	334	338	364	364	376	492	413	419	492	453	265.3	169.4	47.3	-7.9
Arkansas	27	56	54	59	55	52	67	96	81	80	86	88	98	107	123	355.6	92.6	65.4	43.0
Delaware	14	12	4	17	15	18	23	37	32	29	27	27	19	22	22	57.1	28.6	50.0	-18.5
District of Columbia	272	339	329	359	322	361	345	393	421	361	408	291	307	263	221	-18.8	32.7	13.0	-45.8
Florida	171	175	176	205	218	248	238	280	276	316	431	461	503	599	642	275.4	45.0	73.8	49.0
Georgia	351	364	349	286	293	286	291	277	355	281	370	354	347	383	414	17.9	-18.5	29.4	11.9
Kentucky	63	71	73	99	121	119	143	179	155	149	184	153	167	156	196	211.1	88.9	54.6	6.5
Louisiana	204	201	230	211	279	289	260	277	327	313	321	325	349	348	383	87.7	41.7	11.1	19.3
Maryland	338	415	408	456	412	468	430	432	456	456	450	480	502	452	490	45.0	38.5	-3.8	8.9
Mississippi	93	130	133	95	112	106	105	95	97	124	119	108	123	141	167	79.6	14.0	12.3	40.3
North Carolina	272	294	397	402	394	408	402	442	458	472	486	481	476	515	499	83.5	50.0	19.1	2.7
Oklahoma	116	133	106	146	114	135	126	124	129	167	206	159	152	184	171	47.4	16.4	52.6	-17.0
South Carolina	66	93	132	114	124	148	174	141	187	138	179	215	211	215	273	313.6	124.2	20.9	52.5
Tennessee	214	240	204	245	260	234	275	226	230	239	224	246	276	279	285	33.2	9.3	-4.3	27.2
Texas	674	675	850	885	895	889	901	880	874	895	983	1,024	1,096	1,069	967	43.5	31.9	10.6	-1.6
Virginia	164	140	164	177	201	287	275	362	401	406	411	346	429	448	423	157.9	75.0	43.2	2.9
West Virginia	46	68	67	71	81	76	88	98	104	115	82	94	124	81	88	91.3	65.2	7.9	7.3
West	2,280	2,368	2,452	2,655	2,468	2,974	2,927	3,288	3,287	3,152	3,156	3,123	3,156	3,145	3,879	70.1	30.4	6.1	22.9
Alaska	0	0	0	0	0	0	0	9	0	10	11	4	11	2	7	([1])	([1])	([1])	-36.4
Arizona	84	58	68	94	115	110	122	169	146	132	143	149	133	106	170	102.4	31.0	30.0	18.9
California	1,446	1,465	1,582	1,641	1,436	1,879	1,726	1,964	2,165	2,040	1,983	1,859	1,867	1,942	2,417	67.2	29.9	5.5	21.9
Colorado	240	242	213	246	225	233	203	281	225	216	224	249	296	241	253	5.4	-2.9	-3.9	12.9
Hawaii	123	153	174	179	198	191	185	195	151	154	88	132	124	106	115	-6.5	55.3	-53.9	30.7
Idaho	10	22	11	19	10	18	11	25	17	14	24	18	28	21	30	200.0	80.0	33.3	25.0

Table 5-16.—Continued

Region and state	1975-76	1976-77	1977-78	1978-79	1979-80	1980-81	1981-82	1982-83	1983-84	1984-85	1985-86	1986-87	1987-88	1988-89	1989-90	Percent change: 1975-76 to 1989-90	Percent change: 1975-76 to 1980-81	Percent change: 1980-81 to 1985-86	Percent change: 1985-86 to 1989-90
West Continued																			
Montana	17	13	17	7	10	17	22	29	18	20	36	32	33	25	50	194.1	0.0	111.8	38.9
Nevada	8	0	6	3	5	13	12	8	12	12	13	9	8	17	20	150.0	62.5	0.0	53.8
New Mexico	21	24	21	20	21	21	30	40	40	31	46	37	37	40	44	109.5	0.0	119.0	-4.3
Oregon	12	26	28	29	52	55	121	110	118	103	112	107	138	165	194	1516.7	358.3	103.6	73.2
Utah	55	76	61	106	83	102	121	112	97	80	87	139	146	131	132	140.0	85.5	-14.7	51.7
Washington	247	285	254	296	300	325	363	332	287	323	372	370	324	330	427	72.9	31.6	14.5	14.8
Wyoming	17	4	17	15	13	10	11	14	11	17	17	18	11	19	20	17.6	-41.2	70.0	17.6
U.S. Service Schools	—	—	—	—	—	—	—	0	0	0	0	13	—	—	—	(1)	(1)	(1)	(1)
Outlying Areas	179	141	158	152	108	170	130	151	285	142	140	160	137	132	128	-28.5	-5.0	-17.6	-8.6
American Samoa	—	—	—	—	—	—	—	0	0	0	0	—	—	—	—	(1)	(1)	(1)	(1)
Guam	—	—	—	—	—	—	—	0	0	0	0	—	—	—	—	(1)	(1)	(1)	(1)
Northern Marianas	0	0	0	0	—	—	—	—	—	—	—	—	—	—	—	(1)	(1)	(1)	(1)
Puerto Rico	179	141	158	152	108	170	130	151	285	142	140	160	137	132	128	-28.5	-5.0	-17.6	-8.6
Trust Territories	—	—	—	—	—	—	—	—	—	—	—	—	—	—	—	(1)	(1)	(1)	(1)
Virgin Islands	—	—	—	—	—	—	—	0	0	0	—	—	—	—	—	(1)	(1)	(1)	(1)

[1] Insufficient data for calculating a percent change.

—Data not available.

SOURCE: U.S. Department of Education, National Center for Education Statistics, HEGIS, "Degrees and Other Formal Awards Conferred" surveys; and IPEDS, "Completions" surveys.

Table 5-17.—Master's degrees conferred in life sciences, by region and state: 1975-76 to 1989-90

Region and state	1975-76	1976-77	1977-78	1978-79	1979-80	1980-81	1981-82	1982-83	1983-84	1984-85	1985-86	1986-87	1987-88	1988-89	1989-90	Percent change 1975-76 to 1989-90	Percent change 1975-76 to 1980-81	Percent change 1980-81 to 1985-86	Percent change 1985-86 to 1989-90
United States	6,582	7,114	6,806	6,831	6,510	5,978	5,874	5,696	5,406	5,059	5,013	4,952	4,784	4,961	4,861	-26.1	-9.2	-16.1	-3.0
Northeast	1,659	1,835	1,714	1,793	1,690	1,619	1,546	1,466	1,413	1,342	1,409	1,399	1,198	1,357	1,407	-15.2	-2.4	-13.0	-0.1
Connecticut	154	154	128	131	187	158	124	136	141	136	137	132	120	148	155	0.6	2.6	-13.3	13.1
Maine	12	16	24	22	12	18	16	19	17	15	23	14	18	16	21	75.0	50.0	27.8	-8.7
Massachusetts	170	224	186	185	212	198	197	169	157	170	198	166	188	189	196	15.3	16.5	0.0	-1.0
New Hampshire	48	45	38	46	35	26	34	33	26	21	21	19	12	18	18	-62.5	-45.8	-19.2	-14.3
New Jersey	277	277	262	249	231	229	224	201	189	182	183	156	184	240	220	-20.6	-17.3	-20.1	20.2
New York	669	772	727	767	660	697	651	661	638	618	606	710	507	550	638	-4.6	4.2	-13.1	5.3
Pennsylvania	272	286	294	320	293	233	248	203	196	171	200	164	128	144	119	-56.3	-14.3	-14.2	-40.5
Rhode Island	33	50	35	41	31	31	28	20	23	19	18	18	18	25	19	-42.4	-6.1	-41.9	5.6
Vermont	24	11	20	32	29	29	24	24	26	10	23	20	23	27	21	-12.5	20.8	-20.7	-8.7
Midwest	1,865	1,932	1,837	1,784	1,625	1,494	1,511	1,368	1,348	1,289	1,261	1,299	1,235	1,226	1,233	-33.9	-19.9	-15.6	-2.2
Illinois	380	414	373	388	345	319	310	289	284	295	301	286	289	279	274	-27.9	-16.1	-5.6	-9.0
Indiana	177	196	182	137	128	103	131	91	105	104	97	78	102	83	80	-54.8	-41.8	-5.8	-17.5
Iowa	66	81	58	69	59	53	58	53	66	72	60	82	81	57	67	1.5	-19.7	13.2	11.7
Kansas	98	72	114	115	84	105	75	78	85	82	83	73	72	65	67	-31.6	7.1	-21.0	-19.3
Michigan	328	315	288	292	279	262	268	232	225	190	184	196	155	180	201	-38.7	-20.1	-29.8	9.2
Minnesota	134	145	130	112	113	116	121	103	75	75	80	87	73	54	69	-48.5	-13.4	-31.0	-13.8
Missouri	112	112	126	109	104	92	85	84	70	75	68	76	73	86	74	-33.9	-17.9	-26.1	8.8
Nebraska	52	65	76	63	57	38	37	60	50	52	51	49	50	55	40	-23.1	-26.9	34.2	-21.6
North Dakota	25	39	37	40	30	29	27	24	38	25	20	32	28	30	27	8.0	16.0	-31.0	35.0
Ohio	277	294	249	276	266	238	249	240	239	210	218	218	190	221	215	-22.4	-14.1	-8.4	-1.4
South Dakota	23	33	23	23	16	15	24	24	20	20	15	14	14	18	14	-39.1	-34.8	0.0	-6.7
Wisconsin	193	166	181	160	144	124	126	90	91	89	84	108	108	98	105	-45.6	-35.8	-32.3	25.0
South	1,913	2,020	1,997	1,922	1,914	1,723	1,629	1,644	1,505	1,384	1,313	1,334	1,438	1,441	1,322	-30.9	-9.9	-23.8	0.7
Alabama	83	78	84	85	96	90	76	79	76	65	54	56	64	70	70	-15.7	8.4	-40.0	29.6
Arkansas	44	55	52	69	46	46	26	33	32	28	26	33	21	29	26	-40.9	4.5	-43.5	0.0
Delaware	33	25	44	5	8	6	8	26	4	7	2	7	8	5	5	-84.8	-81.8	-66.7	150.0
District of Columbia	139	103	118	82	118	140	129	58	73	65	59	124	135	119	74	-46.8	0.7	-57.9	25.4
Florida	158	156	128	146	152	124	109	120	123	84	87	108	93	96	94	-40.5	-21.5	-29.8	8.0
Georgia	117	133	135	131	135	107	89	72	91	62	70	68	69	82	66	-43.6	-8.5	-34.6	-5.7
Kentucky	83	111	97	93	95	78	68	63	69	59	53	44	34	53	49	-41.0	-6.0	-32.1	-7.5
Louisiana	147	138	120	128	145	110	112	127	84	79	89	72	83	97	70	-52.4	-25.2	-19.1	-21.3
Maryland	73	53	72	73	75	79	76	83	109	94	78	97	88	73	81	11.0	8.2	-1.3	3.8
Mississippi	71	93	93	68	54	55	66	52	46	57	41	37	34	41	29	-59.2	-22.5	-25.5	-29.3
North Carolina	210	185	229	218	215	188	189	208	193	192	169	160	167	182	186	-11.4	-10.5	-10.1	10.1
Oklahoma	69	65	81	60	49	49	60	55	56	53	47	36	43	50	63	-8.7	-29.0	-4.1	34.0
South Carolina	97	88	86	90	100	81	84	79	52	58	55	45	74	45	47	-51.5	-16.5	-32.1	-14.5
Tennessee	118	131	143	130	106	103	83	95	77	103	73	59	80	92	83	-29.7	-12.7	-29.1	13.7
Texas	317	401	296	359	326	276	265	329	258	248	273	254	312	273	247	-22.1	-12.9	-1.1	-9.5
Virginia	106	166	174	144	152	148	142	135	134	114	108	113	113	111	116	9.4	39.6	-27.0	7.4
West Virginia	48	39	45	41	42	43	47	30	28	16	29	21	20	23	16	-66.7	-10.4	-32.6	-44.8
West	1,145	1,327	1,258	1,332	1,281	1,141	1,187	1,218	1,140	1,038	1,018	920	913	937	899	-21.5	-0.3	-10.8	-11.7
Alaska	5	4	6	5	13	6	5	11	16	9	9	7	7	5	7	40.0	20.0	50.0	-22.2
Arizona	65	100	80	108	90	78	99	83	85	77	62	89	101	67	67	3.1	20.0	-20.5	8.1
California	630	657	660	693	663	599	628	661	608	541	553	508	465	460	447	-29.0	-4.9	-7.7	-19.2
Colorado	78	116	138	95	78	97	85	119	87	65	69	51	80	81	77	-1.3	24.4	-28.9	11.6
Hawaii	35	29	42	43	42	35	31	36	31	38	27	39	28	34	28	-20.0	0.0	-22.9	3.7
Idaho	22	37	23	31	40	22	25	21	25	26	30	20	25	23	25	13.6	0.0	36.4	-16.7

Table 5-17.—Continued

Region and state	1975-76	1976-77	1977-78	1978-79	1979-80	1980-81	1981-82	1982-83	1983-84	1984-85	1985-86	1986-87	1987-88	1988-89	1989-90	Percent change 1975-76 to 1989-90	Percent change 1975-76 to 1980-81	Percent change 1980-81 to 1985-86	Percent change 1985-86 to 1989-90
West Continued																			
Montana	39	40	29	36	34	19	25	23	20	26	20	24	13	20	11	-71.8	-51.3	5.3	-45.0
Nevada	16	17	16	4	9	12	14	13	18	10	17	9	6	8	9	-43.8	-25.0	41.7	-47.1
New Mexico ..	38	44	40	51	41	29	19	31	32	26	23	14	26	27	22	-42.1	-23.7	-20.7	-4.3
Oregon	59	66	64	53	69	63	63	65	55	45	42	39	36	49	57	-3.4	6.8	-33.3	35.7
Utah.........	47	57	30	58	66	72	52	57	44	52	49	26	35	36	51	8.5	53.2	-31.9	4.1
Washington...	88	129	110	127	123	88	121	82	90	94	90	78	79	98	82	-6.8	0.0	2.3	-8.9
Wyoming.....	23	31	20	28	13	21	20	16	29	29	27	16	12	29	16	-30.4	-8.7	28.6	-40.7
U.S. Service Schools......	0	0	0	0	0	1	1	0	0	6	12	—	—	—	—	(1)	(1)	1100.0	(1)
Outlying Areas..	39	40	45	48	26	37	57	39	31	32	31	47	41	18	32	-17.9	-5.1	-16.2	3.2
American Samoa.....	—	—	—	—	—	—	—	—	—	—	—	—	—	—	—	(1)	(1)	(1)	(1)
Guam	3	5	3	1	9	3	5	6	2	2	4	3	2	3	2	-33.3	0.0	33.3	-50.0
Northern Marianas...	—	—	—	—	—	—	—	—	—	—	—	—	—	—	—	(1)	(1)	(1)	(1)
Puerto Rico ..	36	35	42	47	17	34	52	33	29	30	27	44	39	15	30	-16.7	-5.6	-20.6	11.1
Trust Territories	—	—	—	—	—	—	—	—	—	—	—	—	—	—	—	(1)	(1)	(1)	(1)
Virgin Islands .	0	0	0	0	0	0	0	0	0	0	0	—	—	—	—	(1)	(1)	(1)	(1)

[1] Insufficient data for calculating a percent change.

—Data not available.

SOURCE: U.S. Department of Education, National Center for Education Statistics, HEGIS, "Degrees and Other Formal Awards Conferred" surveys; and IPEDS, "Completions" surveys.

Table 5-18.—Master's degrees conferred in physical sciences, by region and state: 1975-76 to 1989-90

Region and state	1975-76	1976-77	1977-78	1978-79	1979-80	1980-81	1981-82	1982-83	1983-84	1984-85	1985-86	1986-87	1987-88	1988-89	1989-90	Percent change 1975-76 to 1989-90	Percent change 1975-76 to 1980-81	Percent change 1980-81 to 1985-86	Percent change 1985-86 to 1989-90
United States . . .	5,466	5,331	5,561	5,451	5,219	5,284	5,514	5,290	5,576	5,796	5,902	5,629	5,733	5,723	5,447	-0.3	-3.3	11.7	-7.7
Northeast	1,612	1,414	1,419	1,336	1,361	1,285	1,312	1,277	1,421	1,431	1,438	1,374	1,343	1,420	1,265	-21.5	-20.3	11.9	-12.0
Connecticut . .	93	112	79	81	90	96	58	70	104	106	140	83	83	92	98	5.4	3.2	45.8	-30.0
Maine	8	14	13	19	23	13	8	20	13	11	18	13	14	14	15	87.5	62.5	38.5	-16.7
Massachusetts	223	229	232	209	255	227	224	210	233	210	259	216	238	256	231	3.6	1.8	14.1	-10.8
New Hampshire	22	21	14	17	20	24	17	20	26	19	26	25	15	22	28	27.3	9.1	8.3	7.7
New Jersey . . .	192	176	170	156	153	156	201	164	158	203	171	166	127	147	128	-33.3	-18.8	9.6	-25.1
New York	684	517	566	503	495	469	474	456	480	532	516	549	559	561	504	-26.3	-31.4	10.0	-2.3
Pennsylvania .	311	270	287	299	279	240	265	269	335	283	247	254	250	265	218	-29.9	-22.8	2.9	-11.7
Rhode Island .	69	64	46	45	40	51	61	55	67	58	50	62	52	51	38	-44.9	-26.1	-2.0	-24.0
Vermont	10	11	12	7	6	9	4	13	5	9	11	6	5	12	5	-50.0	-10.0	22.2	-54.5
Midwest	1,394	1,446	1,490	1,494	1,351	1,320	1,435	1,337	1,336	1,430	1,455	1,424	1,382	1,408	1,375	-1.4	-5.3	10.2	-5.5
Illinois	329	348	300	351	286	306	320	279	280	314	313	323	302	309	293	-10.9	-7.0	2.3	-6.4
Indiana	160	146	185	127	131	148	146	121	131	146	139	134	147	128	168	5.0	-7.5	-6.1	20.9
Iowa	55	66	64	69	52	54	46	68	64	55	59	58	65	56	38	-30.9	-1.8	9.3	-35.6
Kansas	55	74	104	75	65	71	63	81	70	74	83	77	79	80	70	27.3	29.1	16.9	-15.7
Michigan	178	153	220	219	167	163	167	186	222	196	205	166	173	196	192	7.9	-8.4	25.8	-6.3
Minnesota	39	56	54	63	53	38	45	46	32	52	37	56	44	50	50	28.2	-2.6	-2.6	35.1
Missouri	86	100	85	101	93	91	117	86	87	98	104	110	102	103	89	3.5	5.8	14.3	-14.4
Nebraska	29	24	24	24	22	14	29	32	29	24	50	35	46	29	37	27.6	-51.7	257.1	-26.0
North Dakota .	9	21	17	21	18	23	8	18	20	26	18	14	22	18	17	88.9	155.6	-21.7	-5.6
Ohio	306	319	318	304	312	304	336	295	269	298	320	330	266	314	294	-3.9	-0.7	5.3	-8.1
South Dakota .	30	24	18	38	23	11	26	18	21	32	18	19	27	20	27	-10.0	-63.3	63.6	50.0
Wisconsin	118	115	101	102	129	97	132	107	111	115	109	102	109	105	100	-15.3	-17.8	12.4	-8.3
South	1,297	1,290	1,340	1,306	1,222	1,327	1,382	1,322	1,359	1,476	1,538	1,426	1,579	1,517	1,438	10.9	2.3	15.9	-6.5
Alabama	35	45	39	28	31	31	44	27	30	57	50	57	49	53	54	54.3	-11.4	61.3	8.0
Arkansas	21	22	27	28	18	25	22	29	22	26	20	23	15	19	18	-14.3	19.0	-20.0	-10.0
Delaware	16	27	27	31	44	32	39	14	34	16	24	15	17	23	22	37.5	100.0	-25.0	-8.3
District of Columbia	37	76	35	40	40	60	46	44	48	44	46	47	67	82	58	56.8	62.2	-23.3	26.1
Florida	144	125	137	116	144	125	141	86	99	109	110	100	146	125	126	-12.5	-13.2	-12.0	14.5
Georgia	84	82	88	83	83	96	112	97	77	85	74	71	92	86	75	-10.7	14.3	-22.9	1.4
Kentucky	55	57	46	58	55	51	48	68	57	53	48	46	39	45	46	-16.4	-7.3	-5.9	-4.2
Louisiana	53	55	52	63	51	57	74	76	75	82	91	109	103	96	95	79.2	7.5	59.6	4.4
Maryland	105	86	108	77	66	76	63	76	91	88	98	87	99	96	116	10.5	-27.6	28.9	18.4
Mississippi . . .	34	10	31	17	15	32	31	21	20	38	46	30	46	33	26	-23.5	-5.9	43.8	-43.5
North Carolina	69	74	98	69	95	92	110	99	91	92	91	111	92	95	77	11.6	33.3	-1.1	-15.4
Oklahoma	95	44	54	83	60	55	46	62	62	102	87	85	98	77	78	-17.9	-42.1	58.2	-10.3
South Carolina	45	39	49	36	33	45	62	45	58	65	62	51	45	51	62	37.8	0.0	37.8	0.0
Tennessee . . .	65	60	55	45	53	51	54	67	51	52	62	65	68	68	64	-1.5	-21.5	21.6	3.2
Texas	313	361	380	394	327	362	365	396	401	447	499	435	473	451	395	26.2	15.7	37.8	-20.8
Virginia	86	98	88	92	77	110	88	92	106	93	97	72	106	94	109	26.7	27.9	-11.8	12.4
West Virginia .	40	29	26	46	30	27	37	23	37	27	33	22	24	23	17	-57.5	-32.5	22.2	-48.5
West	1,101	1,132	1,276	1,270	1,247	1,295	1,342	1,315	1,383	1,422	1,410	1,405	1,429	1,378	1,369	24.3	17.6	8.9	-2.9
Alaska	6	10	6	12	14	13	18	23	20	19	23	20	15	18	18	200.0	116.7	76.9	-21.7
Arizona	98	105	111	111	105	110	120	115	142	124	132	123	164	138	131	33.7	12.2	20.0	-0.8
California	554	470	556	557	541	549	575	577	625	642	544	595	636	556	607	9.6	-0.9	-0.9	11.6
Colorado	90	126	123	152	129	154	145	146	135	176	174	107	145	133	127	41.1	71.1	13.0	-27.0
Hawaii	24	33	23	28	21	21	16	23	26	25	19	31	19	24	25	4.2	-12.5	-9.5	31.6
Idaho	12	24	23	23	39	29	28	19	20	32	30	45	24	22	27	125.0	141.7	3.4	-10.0

Table 5-18.—Continued

Region and state	1975-76	1976-77	1977-78	1978-79	1979-80	1980-81	1981-82	1982-83	1983-84	1984-85	1985-86	1986-87	1987-88	1988-89	1989-90	Percent change: 1975-76 to 1989-90	Percent change: 1975-76 to 1980-81	Percent change: 1980-81 to 1985-86	Percent change: 1985-86 to 1989-90
West Continued																			
Montana	21	18	29	17	30	33	25	36	29	35	42	42	34	32	40	90.5	57.1	27.3	-4.8
Nevada	15	15	26	13	28	18	26	17	13	18	20	24	22	22	31	106.7	20.0	11.1	55.0
New Mexico ..	58	52	65	77	55	81	76	60	86	74	82	94	78	109	65	12.1	39.7	1.2	-20.7
Oregon	72	98	121	86	80	93	98	80	78	80	106	89	85	112	79	9.7	29.2	14.0	-25.5
Utah.........	64	58	69	50	81	71	65	33	47	50	70	64	66	65	64	0.0	10.9	-1.4	-8.6
Washington...	71	94	98	117	102	113	134	151	132	123	141	138	123	123	139	95.8	59.2	24.8	-1.4
Wyoming	16	29	26	27	22	10	16	35	30	24	27	33	18	24	16	0.0	-37.5	170.0	-40.7
U.S. Service Schools......	62	49	36	45	38	57	43	39	77	37	61	0	0	0	0	-100.0	-8.1	7.0	-100.0
Outlying Areas ..	19	14	15	13	14	16	12	11	18	20	14	13	14	27	21	10.5	-15.8	-12.5	50.0
American Samoa.....	—	—	—	—	—	—	—	—	—	—	—	—	—	—	—	(1)	(1)	(1)	(1)
Guam	0	0	0	—	—	0	0	0	0	—	0	—	0	—	—	(1)	(1)	(1)	(1)
Northern Marianas...	—	—	—	—	—	—	—	—	—	—	—	—	—	—	—	(1)	(1)	(1)	(1)
Puerto Rico ..	19	14	15	13	14	16	12	11	18	20	14	13	14	27	21	10.5	-15.8	-12.5	50.0
Trust Territories	—	—	—	—	—	—	—	—	—	—	—	—	—	—	—	(1)	(1)	(1)	(1)
Virgin Islands .	—	0	0	0	0	0	0	0	0	0	0	—	—	—	—	(1)	(1)	(1)	(1)

[1] Insufficient data for calculating a percent change.

—Data not available.

SOURCE: U.S. Department of Education, National Center for Education Statistics, HEGIS, "Degrees and Other Formal Awards Conferred" surveys; and IPEDS, "Completions" surveys.

Table 5-19.—Doctor's degrees conferred in agricultural sciences, by region and state: 1975-76 to 1989-90

Region and state	1975-76	1976-77	1977-78	1978-79	1979-80	1980-81	1981-82	1982-83	1983-84	1984-85	1985-86	1986-87	1987-88	1988-89	1989-90	Percent change: 1975-76 to 1989-90	Percent change: 1975-76 to 1980-81	Percent change: 1980-81 to 1985-86	Percent change: 1985-86 to 1989-90
United States . . .	928	893	971	950	991	1,067	1,079	1,149	1,172	1,213	1,158	1,049	1,142	1,183	1,272	37.1	15.0	8.5	9.8
Northeast	129	115	144	120	105	110	115	117	109	130	120	119	134	148	149	15.5	-14.7	9.1	24.2
Connecticut . .	4	14	4	12	6	5	7	6	5	11	10	10	11	6	11	175.0	25.0	100.0	10.0
Maine	1	3	10	3	1	3	3	2	3	2	4	6	1	4	5	400.0	200.0	33.3	25.0
Massachusetts	27	22	18	17	11	18	18	9	13	21	14	15	9	16	19	-29.6	-33.3	-22.2	35.7
New Hampshire	0	0	0	0	0	0	0	0	0	0	1	2	3	1	—	([1])	([1])	([1])	([1])
New Jersey . . .	13	19	18	13	10	8	9	6	12	23	13	18	14	13	13	0.0	-38.5	62.5	0.0
New York	55	38	74	57	55	55	62	71	54	62	61	42	81	91	86	56.4	0.0	10.9	41.0
Pennsylvania .	27	16	17	18	19	19	15	19	19	8	14	23	11	17	14	-48.1	-29.6	-26.3	0.0
Rhode Island .	0	0	0	0	3	0	0	3	3	2	2	1	3	—	1	([1])	([1])	([1])	-50.0
Vermont	2	3	3	0	0	2	1	1	0	1	1	2	1	—	0	-100.0	0.0	-50.0	-100.0
Midwest	368	355	366	372	378	426	418	439	431	473	420	389	416	404	444	20.7	15.8	-1.4	5.7
Illinois	27	32	36	35	43	32	39	43	34	48	48	53	56	49	51	88.9	18.5	50.0	6.3
Indiana	45	30	30	51	44	38	30	31	46	33	34	33	29	35	34	-24.4	-15.6	-10.5	0.0
Iowa	51	28	39	40	46	62	51	39	47	62	45	35	38	43	33	-35.3	21.6	-27.4	-26.7
Kansas	22	24	24	24	22	29	23	30	16	41	29	34	31	35	37	68.2	31.8	0.0	27.6
Michigan	76	90	77	72	69	89	79	83	73	58	63	61	71	67	75	-1.3	17.1	-29.2	19.0
Minnesota	28	27	31	22	25	41	45	46	43	55	48	44	42	42	56	100.0	46.4	17.1	16.7
Missouri	22	29	21	27	25	17	21	19	28	34	26	21	19	27	42	90.9	-22.7	52.9	61.5
Nebraska	13	17	17	23	28	38	32	37	33	29	35	34	38	24	32	146.2	192.3	-7.9	-8.6
North Dakota .	4	14	4	9	11	8	6	11	12	9	15	10	14	10	7	75.0	100.0	87.5	-53.3
Ohio	41	28	22	32	26	32	46	32	37	39	23	26	23	32	38	-7.3	-22.0	-28.1	65.2
South Dakota .	3	4	4	3	3	2	1	1	4	6	9	7	3	9	8	166.7	-33.3	350.0	-11.1
Wisconsin	36	32	61	34	36	38	45	67	58	59	45	31	52	31	31	-13.9	5.6	18.4	-31.1
South	241	228	261	265	290	268	299	336	362	340	387	329	331	373	411	70.5	11.2	44.4	6.2
Alabama	9	10	13	13	20	10	17	16	14	22	20	15	19	16	21	133.3	11.1	100.0	5.0
Arkansas	5	3	7	7	10	18	14	21	25	25	25	16	8	9	19	280.0	260.0	38.9	-24.0
Delaware	0	0	1	0	0	0	1	0	0	0	3	2	1	1	1	([1])	([1])	([1])	-66.7
District of Columbia . . .	—	0	0	0	0	0	0	0	0	0	0	—	—	—	—	([1])	([1])	([1])	([1])
Florida	31	23	26	34	37	30	27	39	42	32	36	42	38	40	54	74.2	-3.2	20.0	50.0
Georgia	14	25	26	12	10	17	23	22	12	30	24	25	23	35	28	100.0	21.4	41.2	16.7
Kentucky	13	16	10	14	12	18	13	18	18	21	11	22	15	24	11	-15.4	38.5	-38.9	0.0
Louisiana	28	9	12	16	16	12	12	12	17	15	17	9	24	19	26	-7.1	-57.1	41.7	52.9
Maryland	11	12	10	12	20	9	12	10	11	6	13	12	13	18	14	27.3	-18.2	44.4	7.7
Mississippi . . .	15	8	8	7	12	19	17	20	22	19	23	19	27	17	20	33.3	26.7	21.1	-13.0
North Carolina	22	22	35	31	31	29	37	36	52	38	44	44	35	37	44	100.0	31.8	51.7	0.0
Oklahoma	15	15	22	15	21	22	21	18	18	22	34	21	24	22	21	40.0	46.7	54.5	-38.2
South Carolina	2	1	2	2	3	3	4	5	6	3	4	9	12	9	12	500.0	50.0	33.3	200.0
Tennessee . . .	15	13	7	8	10	6	17	8	11	15	6	7	8	13	10	-33.3	-60.0	0.0	66.7
Texas	36	55	64	62	60	47	53	71	79	64	89	62	66	75	86	138.9	30.6	89.4	-3.4
Virginia	23	14	14	28	25	22	26	30	31	26	36	23	15	32	39	69.6	-4.3	63.6	8.3
West Virginia .	2	2	4	4	3	6	5	10	4	2	2	1	3	6	5	150.0	200.0	-66.7	150.0
West	190	195	200	193	218	263	247	257	270	270	231	212	261	258	268	41.1	38.4	-12.2	16.0
Alaska	2	0	0	0	0	0	0	2	0	0	1	—	—	0	—	([1])	-100.0	([1])	([1])
Arizona	10	19	16	12	21	17	18	15	18	24	27	16	20	22	21	110.0	70.0	58.8	-22.2
California	33	41	42	30	42	49	42	53	59	52	46	40	46	63	55	66.7	48.5	-6.1	19.6
Colorado	43	22	22	25	31	31	30	31	40	27	27	20	28	29	33	-23.3	-27.9	-12.9	22.2
Hawaii	12	12	19	8	8	15	11	13	13	21	9	15	13	14	13	8.3	25.0	-40.0	44.4
Idaho	4	7	10	14	3	15	13	8	9	9	9	9	15	8	17	325.0	275.0	-40.0	88.9

Table 5-19.—Continued

Region and state	1975-76	1976-77	1977-78	1978-79	1979-80	1980-81	1981-82	1982-83	1983-84	1984-85	1985-86	1986-87	1987-88	1988-89	1989-90	Percent change 1975-76 to 1989-90	Percent change 1975-76 to 1980-81	Percent change 1980-81 to 1985-86	Percent change 1985-86 to 1989-90
West Continued																			
Montana	6	5	5	3	8	7	4	5	0	5	4	7	4	4	7	16.7	16.7	-42.9	75.0
Nevada	0	0	0	0	0	0	0	0	0	0	0	—	—	—	—	(1)	(1)	(1)	(1)
New Mexico	1	2	3	4	7	8	9	13	10	12	15	12	7	17	20	1900.0	700.0	87.5	33.3
Oregon	26	36	23	30	28	45	42	54	53	59	35	32	45	48	52	100.0	73.1	-22.2	48.6
Utah	16	13	18	19	15	20	23	11	16	14	11	16	21	10	7	-56.3	25.0	-45.0	-36.4
Washington	36	31	37	45	50	47	53	49	45	43	45	40	56	39	37	2.8	30.6	-4.3	-17.8
Wyoming	1	7	5	3	5	9	2	3	7	4	2	5	6	4	6	500.0	800.0	-77.8	200.0
U.S. Service Schools	—	—	—	—	—	—	—	—	—	—	—	—	—	—	—	(1)	(1)	(1)	(1)
Outlying Areas	0	0	0	0	0	0	0	0	0	0	0	0	0	0	0	(1)	(1)	(1)	(1)
American Samoa	—	—	—	—	—	—	—	—	—	—	—	—	—	—	—	(1)	(1)	(1)	(1)
Guam	—	—	—	0	—	0	0	0	0	0	0	—	0	—	—	(1)	(1)	(1)	(1)
Northern Marianas	—	—	—	—	—	—	—	—	—	—	—	—	—	—	—	(1)	(1)	(1)	(1)
Puerto Rico	0	0	0	0	0	0	0	0	0	0	0	—	—	0	0	(1)	(1)	(1)	(1)
Trust Territories	—	—	—	—	—	—	—	0	0	0	0	—	—	—	—	(1)	(1)	(1)	(1)
Virgin Islands	—	—	—	—	—	—	—	0	0	0	0	—	—	—	—	(1)	(1)	(1)	(1)

[1] Insufficient data for calculating a percent change.

—Data not available.

SOURCE: U.S. Department of Education, National Center for Education Statistics, HEGIS, "Degrees and Other Formal Awards Conferred" surveys; and IPEDS, "Completions" surveys.

Table 5-20.—Doctor's degrees conferred in computer sciences, by region and state: 1975-76 to 1989-90

Region and state	1975-76	1976-77	1977-78	1978-79	1979-80	1980-81	1981-82	1982-83	1983-84	1984-85	1985-86	1986-87	1987-88	1988-89	1989-90	Percent change			
																1975-76 to 1989-90	1975-76 to 1980-81	1980-81 to 1985-86	1985-86 to 1989-90
United States	244	216	196	236	240	252	251	262	251	248	344	374	428	551	623	155.3	3.3	36.5	81.1
Northeast	60	70	55	65	63	69	74	68	72	82	90	104	119	155	153	155.0	15.0	30.4	70.0
Connecticut	10	2	9	7	8	7	5	4	2	9	5	9	5	9	13	30.0	-30.0	-28.6	160.0
Maine	—	0	0	0	0	0	0	0	0	0	0	—	—	—	—	([1])	([1])	([1])	([1])
Massachusetts	1	5	4	4	4	5	4	7	6	8	12	11	14	16	13	1200.0	400.0	140.0	8.3
New Hampshire	0	0	0	0	0	0	0	0	0	0	0	—	—	—	0	([1])	([1])	([1])	([1])
New Jersey	0	2	2	1	1	4	3	6	3	6	4	3	4	9	12	([1])	([1])	0.0	200.0
New York	26	34	13	28	32	26	27	24	30	26	36	55	51	74	69	165.4	0.0	38.5	91.7
Pennsylvania	23	26	27	24	16	26	32	27	28	31	31	23	43	43	44	91.3	13.0	19.2	41.9
Rhode Island	0	1	0	1	2	1	3	0	3	2	2	3	2	4	2	([1])	([1])	100.0	0.0
Vermont	—	0	0	0	0	0	0	0	0	0	0	0	—	—	—	([1])	([1])	([1])	([1])
Midwest	89	72	60	76	83	66	72	70	81	67	95	110	114	140	183	105.6	-25.8	43.9	92.6
Illinois	23	31	13	22	23	28	24	24	33	22	34	45	54	52	73	217.4	21.7	21.4	114.7
Indiana	8	5	3	8	5	2	4	10	7	4	9	8	11	8	10	25.0	-75.0	350.0	11.1
Iowa	7	4	5	6	7	4	7	4	6	2	4	6	6	5	6	-14.3	-42.9	0.0	50.0
Kansas	2	0	7	3	2	3	7	2	3	3	2	4	3	7	2	0.0	50.0	-33.3	0.0
Michigan	16	13	7	9	14	12	7	8	5	8	11	11	6	12	15	-6.3	-25.0	-8.3	36.4
Minnesota	4	1	1	4	3	4	4	8	7	5	9	5	5	9	15	275.0	0.0	125.0	66.7
Missouri	1	4	3	4	10	3	1	2	2	2	1	6	3	7	9	800.0	200.0	-66.7	800.0
Nebraska	0	0	1	0	0	0	0	0	0	0	0	—	—	4	5	([1])	([1])	([1])	([1])
North Dakota	0	0	0	0	0	0	0	0	0	0	0	—	—	—	2	([1])	([1])	([1])	([1])
Ohio	16	8	10	9	11	7	11	4	9	13	10	12	11	23	33	106.3	-56.3	42.9	230.0
South Dakota	0	0	0	0	0	0	0	0	0	0	0	—	—	—	—	([1])	([1])	([1])	([1])
Wisconsin	12	6	10	11	8	3	7	8	9	8	15	13	15	13	13	8.3	-75.0	400.0	-13.3
South	47	32	37	46	32	45	48	53	43	43	76	81	109	140	157	234.0	-4.3	68.9	106.6
Alabama	0	0	1	1	0	0	0	1	0	2	3	3	2	10	6	([1])	([1])	([1])	100.0
Arkansas	0	0	0	0	0	0	0	0	0	0	0	0	—	—	—	([1])	([1])	([1])	([1])
Delaware	4	1	0	0	1	0	1	0	0	0	3	0	3	4	4	0.0	-100.0	([1])	33.3
District of Columbia	7	4	2	4	2	4	4	3	3	2	10	4	6	6	6	-14.3	-42.9	150.0	-40.0
Florida	0	0	0	0	0	1	0	0	1	1	3	5	17	24	26	([1])	([1])	200.0	766.7
Georgia	4	1	2	1	0	3	4	3	3	2	3	9	8	9	10	150.0	-25.0	0.0	233.3
Kentucky	0	0	0	0	0	0	0	0	0	0	0	0	0	0	2	([1])	([1])	([1])	([1])
Louisiana	4	3	1	6	1	3	3	6	3	2	4	5	12	16	13	225.0	-25.0	33.3	225.0
Maryland	7	6	6	8	7	9	9	6	10	9	12	9	11	13	30	328.6	28.6	33.3	150.0
Mississippi	0	0	0	0	0	0	0	0	0	0	0	0	0	0	0	([1])	([1])	([1])	([1])
North Carolina	3	4	10	4	5	6	4	5	5	3	7	9	7	10	15	400.0	100.0	16.7	114.3
Oklahoma	1	0	1	1	1	1	1	3	3	1	4	3	4	5	1	0.0	0.0	300.0	-75.0
South Carolina	0	0	0	0	0	0	0	0	0	0	0	2	1	2	5	([1])	([1])	([1])	([1])
Tennessee	1	2	1	2	4	1	1	1	3	1	2	1	4	3	4	300.0	0.0	100.0	100.0
Texas	15	9	12	15	9	17	19	19	8	13	18	26	27	32	29	93.3	13.3	5.9	61.1
Virginia	1	2	1	4	2	0	2	6	4	7	7	5	7	6	6	500.0	-100.0	([1])	-14.3
West Virginia	0	0	0	0	0	0	0	0	0	0	0	0	0	0	0	([1])	([1])	([1])	([1])
West	48	42	44	49	62	72	57	71	55	56	83	79	86	116	130	170.8	50.0	15.3	56.6
Alaska	0	—	0	0	—	—	0	0	0	0	0	—	—	—	—	([1])	([1])	([1])	([1])
Arizona	0	1	1	0	1	0	2	0	0	0	2	2	2	5	8	([1])	([1])	([1])	300.0
California	36	35	33	36	50	63	41	58	45	49	72	57	66	85	91	152.8	75.0	14.3	26.4
Colorado	0	2	0	0	3	3	3	1	2	1	1	7	2	1	8	([1])	([1])	-66.7	700.0
Hawaii	0	0	0	0	0	0	0	0	0	0	0	0	—	—	—	([1])	([1])	([1])	([1])
Idaho	0	0	0	0	0	0	0	0	0	0	0	—	—	—	—	([1])	([1])	([1])	([1])

Table 5-20.—Continued

Region and state	1975-76	1976-77	1977-78	1978-79	1979-80	1980-81	1981-82	1982-83	1983-84	1984-85	1985-86	1986-87	1987-88	1988-89	1989-90	Percent change 1975-76 to 1989-90	Percent change 1975-76 to 1980-81	Percent change 1980-81 to 1985-86	Percent change 1985-86 to 1989-90
West Continued																			
Montana	0	0	0	0	0	0	0	0	0	0	0	—	—	—	0	(1)	(1)	(1)	(1)
Nevada	—	—	—	—	0	0	0	0	0	0	0	—	—	—	—	(1)	(1)	(1)	(1)
New Mexico	0	0	0	0	0	0	0	1	0	0	1	—	3	4	3	(1)	(1)	(1)	200.0
Oregon	0	0	2	1	0	0	2	2	0	1	2	2	2	9	4	(1)	(1)	(1)	100.0
Utah	7	2	5	5	3	2	4	0	0	0	0	0	—	—	—	(1)	-71.4	-100.0	(1)
Washington	5	2	3	7	5	4	5	9	8	5	5	11	11	12	16	220.0	-20.0	25.0	220.0
Wyoming	0	0	0	0	0	0	0	0	0	0	0	—	—	—	—	(1)	(1)	(1)	(1)
U.S. Service Schools	0	0	0	0	0	0	0	0	0	0	0	—	—	—	—	(1)	(1)	(1)	(1)
Outlying Areas	0	0	0	0	0	0	0	0	0	0	0	0	0	0	0	(1)	(1)	(1)	(1)
American Samoa	—	—	—	—	—	—	—	—	—	—	—	—	—	—	—	(1)	(1)	(1)	(1)
Guam	—	—	—	—	—	—	—	0	0	0	0	—	—	—	—	(1)	(1)	(1)	(1)
Northern Marianas	—	—	—	—	—	—	—	—	—	—	—	—	—	—	—	(1)	(1)	(1)	(1)
Puerto Rico	0	0	0	0	0	0	0	0	0	0	0	—	—	0	0	(1)	(1)	(1)	(1)
Trust Territories	—	—	—	—	—	—	—	—	—	—	—	—	—	—	—	(1)	(1)	(1)	(1)
Virgin Islands	—	—	—	—	—	—	—	—	—	0	—	—	—	—	—	(1)	(1)	(1)	(1)

[1] Insufficient data for calculating a percent change.

—Data not available.

SOURCE: U.S. Department of Education, National Center for Education Statistics, HEGIS, "Degrees and Other Formal Awards Conferred" surveys; and IPEDS, "Completions" surveys.

Table 5-21.—Doctor's degrees conferred in engineering, by region and state: 1975-76 to 1989-90

Region and state	1975-76	1976-77	1977-78	1978-79	1979-80	1980-81	1981-82	1982-83	1983-84	1984-85	1985-86	1986-87	1987-88	1988-89	1989-90	Percent change: 1975-76 to 1989-90	Percent change: 1975-76 to 1980-81	Percent change: 1980-81 to 1985-86	Percent change: 1985-86 to 1989-90
United States . . .	2,821	2,586	2,440	2,506	2,507	2,561	2,636	2,831	2,981	3,230	3,410	3,818	4,191	4,523	4,965	76.0	-9.2	33.2	45.6
Northeast	775	691	697	689	671	704	758	777	820	899	906	1,038	1,113	1,207	1,297	67.4	-9.2	28.7	43.2
Connecticut . .	23	29	27	16	19	28	18	35	38	37	27	50	45	49	36	56.5	21.7	-3.6	33.3
Maine	1	6	0	0	0	0	1	1	0	2	4	3	1	4	2	100.0	-100.0	(1)	-50.0
Massachusetts	218	189	226	202	195	209	221	232	217	247	253	291	319	315	353	61.9	-4.1	21.1	39.5
New Hampshire	7	4	5	5	9	9	3	8	4	4	6	6	8	3	8	14.3	28.6	-33.3	33.3
New Jersey . . .	66	74	61	60	56	65	71	71	69	77	68	95	93	109	122	84.8	-1.5	4.6	79.4
New York	263	231	234	236	235	241	254	255	272	296	319	357	341	384	411	56.3	-8.4	32.4	28.8
Pennsylvania .	169	139	129	156	134	138	164	158	204	219	213	215	272	305	333	97.0	-18.3	54.3	56.3
Rhode Island .	24	18	14	13	22	14	24	15	16	15	14	20	30	33	27	12.5	-41.7	0.0	92.9
Vermont	4	1	1	1	1	0	2	2	0	2	2	1	4	5	5	25.0	-100.0	(1)	150.0
Midwest	720	649	630	710	710	742	731	791	782	900	952	1,063	1,130	1,218	1,321	83.5	3.1	28.3	38.8
Illinois	171	170	144	188	185	199	180	184	180	221	211	224	267	265	301	76.0	16.4	6.0	42.7
Indiana	95	94	91	91	103	110	105	94	112	112	130	110	131	157	176	85.3	15.8	18.2	35.4
Iowa	53	44	37	51	49	42	58	59	47	58	69	97	97	78	84	58.5	-20.8	64.3	21.7
Kansas	20	26	25	17	25	28	25	27	23	32	23	21	29	39	40	100.0	40.0	-17.9	73.9
Michigan	91	70	86	82	88	91	111	120	115	132	124	167	165	198	234	157.1	0.0	36.3	88.7
Minnesota	38	28	37	39	37	46	38	37	40	59	71	64	65	83	93	144.7	21.1	54.3	31.0
Missouri	50	72	44	56	50	50	32	54	42	43	54	54	64	68	73	46.0	0.0	8.0	35.2
Nebraska	7	3	3	6	2	7	9	8	15	8	6	8	9	9	17	142.9	0.0	-14.3	183.3
North Dakota .	0	0	0	0	0	0	0	0	0	0	3	1	4	3	4	(1)	(1)	(1)	33.3
Ohio	127	90	98	117	117	101	107	148	132	140	159	214	172	223	197	55.1	-20.5	57.4	23.9
South Dakota .	2	1	0	0	1	1	0	0	2	0	1	3	6	2	4	100.0	-50.0	0.0	300.0
Wisconsin	66	51	65	63	53	67	66	60	74	95	101	100	121	93	98	48.5	1.5	50.7	-3.0
South	604	603	509	513	520	479	509	583	639	680	764	851	985	1,092	1,210	100.3	-20.7	59.5	58.4
Alabama	10	14	11	13	22	12	19	11	5	18	21	31	36	44	41	310.0	20.0	75.0	95.2
Arkansas	16	7	2	3	5	5	7	4	12	7	16	6	8	6	8	-50.0	-68.8	220.0	-50.0
Delaware	12	7	13	10	15	11	24	16	0	23	22	25	38	26	26	116.7	-8.3	100.0	18.2
District of Columbia . . .	24	24	30	10	20	15	26	24	18	17	34	20	32	33	43	79.2	-37.5	126.7	26.5
Florida	46	56	45	42	25	33	29	38	56	55	68	76	103	85	114	147.8	-28.3	106.1	67.6
Georgia	35	38	38	27	35	30	31	51	54	35	54	45	63	81	81	131.4	-14.3	80.0	50.0
Kentucky	12	11	8	12	12	10	7	9	9	7	11	7	11	18	24	100.0	-16.7	10.0	118.2
Louisiana	15	19	9	5	9	8	11	16	9	18	20	23	33	30	51	240.0	-46.7	150.0	155.0
Maryland	37	48	45	36	30	29	34	26	36	49	48	55	60	83	84	127.0	-21.6	65.5	75.0
Mississippi . . .	12	9	8	10	9	10	9	12	12	7	14	18	6	16	12	0.0	-16.7	40.0	-14.3
North Carolina	50	38	31	46	28	34	45	41	43	50	53	59	80	90	90	80.0	-32.0	55.9	69.8
Oklahoma	55	43	39	48	58	39	35	41	41	47	43	45	57	52	63	14.5	-29.1	10.3	46.5
South Carolina	21	13	13	14	9	10	11	22	17	23	12	22	19	22	26	23.8	-52.4	20.0	116.7
Tennessee . . .	53	52	24	26	44	29	24	33	46	43	35	35	42	42	72	35.8	-45.3	20.7	105.7
Texas	142	158	131	130	135	132	129	155	185	177	213	262	247	316	329	131.7	-7.0	61.4	54.5
Virginia	52	57	53	64	55	65	60	74	83	90	81	108	132	131	129	148.1	25.0	24.6	59.3
West Virginia .	12	9	9	17	9	7	8	10	13	14	19	14	18	17	17	41.7	-41.7	171.4	-10.5
West	696	631	597	579	591	626	628	674	728	745	779	856	963	1,006	1,137	63.4	-10.1	24.4	46.0
Alaska	0	0	0	0	0	0	0	0	0	0	0	—	0	0	—	(1)	(1)	(1)	(1)
Arizona	24	28	21	19	26	23	35	24	27	32	31	45	52	79	87	262.5	-4.2	34.8	180.6
California	496	457	426	392	416	451	446	473	520	505	537	538	616	642	701	41.3	-9.1	19.1	30.5
Colorado	62	48	46	52	41	55	46	52	50	58	58	84	85	77	105	69.4	-11.3	5.5	81.0
Hawaii	5	7	4	2	2	4	5	0	3	5	4	5	3	2	2	-60.0	-20.0	0.0	-50.0
Idaho	2	3	8	5	5	6	3	5	5	6	2	7	7	15	14	600.0	200.0	-66.7	600.0

Table 5-21.—Continued

Region and state	1975-76	1976-77	1977-78	1978-79	1979-80	1980-81	1981-82	1982-83	1983-84	1984-85	1985-86	1986-87	1987-88	1988-89	1989-90	Percent change 1975-76 to 1989-90	Percent change 1975-76 to 1980-81	Percent change 1980-81 to 1985-86	Percent change 1985-86 to 1989-90
West Continued																			
Montana	3	1	5	3	0	0	3	3	3	1	4	4	4	2	3	0.0	-100.0	([1])	-25.0
Nevada	0	0	0	1	0	0	1	0	1	2	1	2	1	2	3	([1])	([1])	([1])	200.0
New Mexico ..	12	15	9	18	13	8	11	19	14	24	29	28	36	27	41	241.7	-33.3	262.5	41.4
Oregon	13	10	12	14	7	15	15	20	6	18	13	27	38	29	27	107.7	15.4	-13.3	107.7
Utah.........	37	22	30	35	38	29	29	41	58	53	44	44	52	69	71	91.9	-21.6	51.7	61.4
Washington...	40	38	31	36	37	31	33	37	39	36	51	67	63	55	79	97.5	-22.5	64.5	54.9
Wyoming.....	2	2	5	2	6	4	1	0	2	5	5	5	6	7	4	100.0	100.0	25.0	-20.0
U.S. Service Schools......	26	12	7	15	15	10	10	6	12	6	9	10	0	0	0	-100.0	-61.5	-10.0	-100.0
Outlying Areas..	0	0	0	0	0	0	0	0	0	0	0	0	0	0	0	([1])	([1])	([1])	([1])
American Samoa.....	—	—	—	—	—	—	—	—	—	—	—	—	—	—	—	([1])	([1])	([1])	([1])
Guam	—	0	—	0	—	0	0	0	0	0	0	—	—	—	—	([1])	([1])	([1])	([1])
Northern Marianas...	—	—	—	—	—	—	—	—	—	—	—	—	—	—	—	([1])	([1])	([1])	([1])
Puerto Rico ..	0	0	0	0	0	0	0	0	0	0	0	—	—	0	0	([1])	([1])	([1])	([1])
Trust Territories	—	—	—	—	—	—	—	0	0	0	0	—	—	—	—	([1])	([1])	([1])	([1])
Virgin Islands .	—	0	0	—	—	—	—	—	—	—	—	—	—	—	—	([1])	([1])	([1])	([1])

[1] Insufficient data for calculating a percent change.

—Data not available.

SOURCE: U.S. Department of Education, National Center for Education Statistics, HEGIS, "Degrees and Other Formal Awards Conferred" surveys; and IPEDS, "Completions" surveys.

Table 5-22.—Doctor's degrees conferred in health sciences, by region and state:1975-76 to 1989-90

Region and state	1975-76	1976-77	1977-78	1978-79	1979-80	1980-81	1981-82	1982-83	1983-84	1984-85	1985-86	1986-87	1987-88	1988-89	1989-90	Percent change			
																1975-76 to 1989-90	1975-76 to 1980-81	1980-81 to 1985-86	1985-86 to 1989-90
United States . . .	577	538	638	705	771	827	910	1,155	1,163	1,199	1,241	1,213	1,261	1,436	1,543	167.4	43.3	50.1	24.3
Northeast	113	153	186	188	251	231	256	289	272	293	319	288	270	333	315	178.8	104.4	38.1	-1.3
Connecticut . .	6	13	8	3	5	15	12	13	20	14	27	18	19	17	18	200.0	150.0	80.0	-33.3
Maine	0	0	0	0	0	0	0	0	0	0	0	—	—	—	—	([1])	([1])	([1])	([1])
Massachusetts	22	31	41	46	53	52	49	90	91	81	90	61	62	86	76	245.5	136.4	73.1	-15.6
New Hampshire	0	0	0	0	0	0	0	0	0	0	0	—	—	—	—	([1])	([1])	([1])	([1])
New Jersey . . .	2	2	3	4	0	2	3	16	0	0	5	1	0	0	8	300.0	0.0	150.0	60.0
New York	45	61	91	78	118	105	104	89	100	123	110	126	98	132	130	188.9	133.3	4.8	18.2
Pennsylvania .	34	44	43	57	71	55	88	78	61	75	84	82	89	96	77	126.5	61.8	52.7	-8.3
Rhode Island .	4	2	0	0	4	2	0	3	0	0	3	0	2	2	6	50.0	-50.0	50.0	100.0
Vermont	0	0	0	0	0	0	0	0	0	0	0	0	—	0	0	([1])	([1])	([1])	([1])
Midwest	194	152	192	204	221	251	269	242	334	340	303	342	336	348	339	74.7	29.4	20.7	11.9
Illinois	26	19	15	18	25	39	42	48	58	65	68	67	63	62	64	146.2	50.0	74.4	-5.9
Indiana	24	14	21	32	28	23	53	31	44	47	30	57	47	41	44	83.3	-4.2	30.4	46.7
Iowa	24	13	19	25	17	29	28	13	16	17	12	20	16	23	20	-16.7	20.8	-58.6	66.7
Kansas	8	10	7	5	7	13	11	11	7	5	6	11	14	30	13	62.5	62.5	-53.8	116.7
Michigan	24	23	20	35	30	29	33	33	40	33	42	26	41	25	46	91.7	20.8	44.8	9.5
Minnesota	20	18	28	28	28	30	24	22	34	30	33	26	41	38	43	115.0	50.0	10.0	30.3
Missouri	1	2	2	2	8	4	3	3	15	12	11	11	12	18	15	1400.0	300.0	175.0	36.4
Nebraska	2	0	0	0	0	0	0	0	0	0	0	—	0	0	—	([1])	-100.0	([1])	([1])
North Dakota .	0	0	0	0	0	0	0	0	0	0	0	2	3	—	—	([1])	([1])	([1])	([1])
Ohio	44	41	55	40	44	55	45	60	95	104	76	81	68	70	53	20.5	25.0	38.2	-30.3
South Dakota .	0	0	0	0	0	0	0	0	0	0	0	—	—	—	0	([1])	([1])	([1])	([1])
Wisconsin	21	12	25	19	34	29	30	21	25	27	25	41	31	41	41	95.2	38.1	-13.8	64.0
South	179	161	181	241	226	246	258	411	355	377	417	412	474	503	471	163.1	37.4	69.5	12.9
Alabama	0	0	0	2	6	10	14	22	27	21	20	30	33	44	43	([1])	([1])	100.0	115.0
Arkansas	0	0	0	0	0	0	0	130	0	0	0	0	—	0	0	([1])	([1])	([1])	([1])
Delaware	0	0	0	0	0	0	0	0	0	0	0	0	0	0	0	([1])	([1])	([1])	([1])
District of Columbia . . .	9	7	9	7	16	10	11	14	16	11	20	17	13	18	15	66.7	11.1	100.0	-25.0
Florida	5	7	3	6	22	23	30	13	28	28	20	19	33	48	36	620.0	360.0	-13.0	80.0
Georgia	6	5	6	6	7	7	7	11	9	23	15	13	17	16	27	350.0	16.7	114.3	80.0
Kentucky	8	5	0	5	6	7	1	13	21	13	11	21	15	28	15	87.5	-12.5	57.1	36.4
Louisiana	7	3	4	22	10	4	2	1	15	15	31	20	22	30	22	214.3	-42.9	675.0	-29.0
Maryland	50	43	55	47	52	57	51	54	71	82	72	68	93	79	81	62.0	14.0	26.3	12.5
Mississippi . . .	8	3	2	5	4	7	7	2	3	6	1	4	3	5	4	-50.0	-12.5	-85.7	300.0
North Carolina	5	14	22	18	18	22	23	27	31	23	21	27	32	31	32	540.0	340.0	-4.5	52.4
Oklahoma	28	19	14	41	16	9	19	12	12	16	27	12	11	7	12	-57.1	-67.9	200.0	-55.6
South Carolina	0	0	0	0	0	0	0	0	0	0	6	10	5	10	12	([1])	([1])	([1])	100.0
Tennessee . . .	4	5	8	13	2	9	5	7	5	8	8	7	8	5	7	75.0	125.0	-11.1	-12.5
Texas	47	50	55	63	67	77	81	92	96	112	140	136	154	144	135	187.2	63.8	81.8	-3.6
Virginia	2	0	3	6	0	4	7	3	11	10	11	14	22	22	23	1050.0	100.0	175.0	109.1
West Virginia .	0	0	0	0	0	0	0	10	10	9	14	14	13	16	7	([1])	([1])	([1])	-50.0
West	91	72	79	72	73	99	127	213	202	189	202	171	181	252	418	359.3	8.8	104.0	106.9
Alaska	0	0	0	0	0	0	0	0	0	0	0	—	—	—	—	([1])	([1])	([1])	([1])
Arizona	0	1	0	2	1	4	4	10	12	18	18	17	15	18	8	([1])	([1])	350.0	-55.6
California	61	52	57	44	47	63	74	102	95	88	93	75	93	146	326	434.4	3.3	47.6	250.5
Colorado	8	7	8	10	2	7	15	12	11	7	7	6	10	13	16	100.0	-12.5	0.0	128.6
Hawaii	0	0	0	0	0	0	0	2	2	1	3	9	5	6	3	([1])	([1])	([1])	0.0
Idaho	0	0	0	0	0	0	0	0	0	0	0	—	—	—	—	([1])	([1])	([1])	([1])

Table 5-22.—Continued

Region and state	1975-76	1976-77	1977-78	1978-79	1979-80	1980-81	1981-82	1982-83	1983-84	1984-85	1985-86	1986-87	1987-88	1988-89	1989-90	Percent change: 1975-76 to 1989-90	Percent change: 1975-76 to 1980-81	Percent change: 1980-81 to 1985-86	Percent change: 1985-86 to 1989-90
West Continued																			
Montana	0	0	0	0	0	0	1	0	0	0	0	—	1	0	0	([1])	([1])	([1])	([1])
Nevada	0	0	0	0	0	0	0	0	0	0	0	—	—	1	—	([1])	([1])	([1])	([1])
New Mexico ..	0	0	0	0	0	0	0	10	1	3	6	2	9	4	7	([1])	([1])	([1])	16.7
Oregon	1	0	0	0	2	3	4	43	38	37	44	2	2	3	10	900.0	200.0	1366.7	-77.3
Utah.........	3	2	5	0	6	7	10	8	11	7	10	25	14	26	24	700.0	133.3	42.9	140.0
Washington...	18	10	9	16	15	15	19	26	32	28	21	35	32	35	24	33.3	-16.7	40.0	14.3
Wyoming	0	0	0	0	0	0	0	0	0	0	0	—	—	—	—	([1])	([1])	([1])	([1])
U.S. Service Schools......	—	—	—	—	—	—	—	0	0	0	0	—	—	—	—	([1])	([1])	([1])	([1])
Outlying Areas ..	0	0	0	0	0	0	0	3	4	0	0	0	7	1	1	([1])	([1])	([1])	([1])
American Samoa	—	—	—	—	—	—	—	0	0	0	0	—	—	—	—	([1])	([1])	([1])	([1])
Guam	—	—	—	—	—	—	—	0	0	0	0	—	—	—	—	([1])	([1])	([1])	([1])
Northern Marianas ...	0	0	0	0	—	—	—	—	—	—	—	—	—	—	—	([1])	([1])	([1])	([1])
Puerto Rico ..	0	0	0	0	0	0	0	3	4	0	0	—	7	1	1	([1])	([1])	([1])	([1])
Trust Territories	—	—	—	—	—	—	—	—	—	—	—	—	—	—	—	([1])	([1])	([1])	([1])
Virgin Islands .	—	—	—	—	—	—	—	0	0	0	—	—	—	—	—	([1])	([1])	([1])	([1])

[1] Insufficient data for calculating a percent change.

—Data not available.

SOURCE: U.S. Department of Education, National Center for Education Statistics, HEGIS, "Degrees and Other Formal Awards Conferred" surveys; and IPEDS, "Completions" surveys.

Table 5-23.—Doctor's degrees conferred in life sciences, by region and state: 1975-76 to 1989-90

Region and state	1975-76	1976-77	1977-78	1978-79	1979-80	1980-81	1981-82	1982-83	1983-84	1984-85	1985-86	1986-87	1987-88	1988-89	1989-90	Percent change: 1975-76 to 1989-90	Percent change: 1975-76 to 1980-81	Percent change: 1980-81 to 1985-86	Percent change: 1985-86 to 1989-90
United States . . .	3,392	3,397	3,309	3,542	3,636	3,718	3,743	3,341	3,437	3,432	3,358	3,419	3,629	3,520	3,844	13.3	9.6	-9.7	14.5
Northeast	903	880	839	942	917	931	976	820	879	816	836	848	938	883	1,023	13.3	3.1	-10.2	22.4
Connecticut . .	60	66	63	61	63	81	75	67	87	70	66	85	91	62	89	48.3	35.0	-18.5	34.8
Maine	8	4	4	3	1	3	2	3	1	4	3	—	3	1	7	-12.5	-62.5	0.0	133.3
Massachusetts	181	164	153	164	208	172	185	116	168	140	153	195	204	200	215	18.8	-5.0	-11.0	40.5
New Hampshire	23	14	15	20	12	24	23	21	20	12	14	16	13	15	23	0.0	4.3	-41.7	64.3
New Jersey . . .	66	83	73	78	75	92	82	85	71	82	88	72	148	75	107	62.1	39.4	-4.3	21.6
New York	384	390	363	430	380	401	432	369	382	361	378	384	361	375	424	10.4	4.4	-5.7	12.2
Pennsylvania .	157	132	142	151	153	132	147	133	121	121	104	71	96	123	124	-21.0	-15.9	-21.2	19.2
Rhode Island .	18	18	19	23	17	15	22	16	17	20	17	16	13	16	19	5.6	-16.7	13.3	11.8
Vermont	6	9	7	12	8	11	8	10	12	6	13	9	9	16	15	150.0	83.3	18.2	15.4
Midwest	919	923	871	964	995	948	972	795	834	906	807	841	895	878	947	3.0	3.2	-14.9	17.3
Illinois	176	190	178	172	205	205	189	152	172	167	171	162	200	179	213	21.0	16.5	-16.6	24.6
Indiana	114	110	122	87	84	97	77	78	81	83	90	80	81	76	77	-32.5	-14.9	-7.2	-14.4
Iowa	60	48	51	57	50	52	57	48	58	68	63	85	89	60	90	50.0	-13.3	21.2	42.9
Kansas	54	40	60	75	62	46	39	42	49	67	40	42	45	48	42	-22.2	-14.8	-13.0	5.0
Michigan	121	150	119	155	164	128	168	150	135	143	121	119	127	129	125	3.3	5.8	-5.5	3.3
Minnesota	70	74	56	69	86	91	60	58	75	79	73	66	83	67	72	2.9	30.0	-19.8	-1.4
Missouri	40	55	53	77	60	67	80	50	50	54	52	46	53	66	69	72.5	67.5	-22.4	32.7
Nebraska	24	26	15	18	29	28	24	34	17	26	14	24	19	21	21	-12.5	16.7	-50.0	50.0
North Dakota .	13	11	9	22	21	15	11	10	5	14	11	13	16	9	13	0.0	15.4	-26.7	18.2
Ohio	110	108	107	99	113	117	125	94	106	97	97	102	80	119	128	16.4	6.4	-17.1	32.0
South Dakota .	1	5	5	3	2	1	4	2	1	5	0	4	3	4	0	-100.0	0.0	-100.0	(1)
Wisconsin	136	106	96	130	119	101	138	77	85	103	75	98	99	100	97	-28.7	-25.7	-25.7	29.3
South	842	839	876	906	903	994	978	901	917	930	873	946	997	964	1,088	29.2	18.1	-12.2	24.6
Alabama	20	40	26	28	30	26	23	27	43	30	22	27	24	28	25	25.0	30.0	-15.4	13.6
Arkansas	13	7	11	6	9	12	12	12	13	8	16	12	8	15	21	61.5	-7.7	33.3	31.3
Delaware	5	5	12	6	3	7	6	12	0	4	1	7	6	3	3	-40.0	40.0	-85.7	200.0
District of Columbia . . .	39	38	50	67	46	72	54	35	38	56	43	41	48	42	47	20.5	84.6	-40.3	9.3
Florida	83	96	70	91	85	63	54	67	43	63	58	64	65	50	64	-22.9	-24.1	-7.9	10.3
Georgia	51	62	73	61	44	76	68	62	63	70	64	54	81	104	67	31.4	49.0	-15.8	4.7
Kentucky	14	18	20	27	37	33	33	18	27	25	22	29	28	25	19	35.7	135.7	-33.3	-13.6
Louisiana	44	38	25	50	40	35	49	34	27	21	31	43	34	41	52	18.2	-20.5	-11.4	67.7
Maryland	62	68	75	64	56	60	74	89	81	82	91	101	96	99	133	114.5	-3.2	51.7	46.2
Mississippi . . .	38	37	26	23	22	33	33	16	22	17	16	20	25	24	31	-18.4	-13.2	-51.5	93.8
North Carolina	133	106	134	136	155	143	174	194	177	144	154	170	166	146	185	39.1	7.5	7.7	20.1
Oklahoma	42	36	39	38	27	32	33	35	35	33	17	26	29	25	33	-21.4	-23.8	-46.9	94.1
South Carolina	25	19	22	40	28	31	35	31	33	34	35	31	44	31	29	16.0	24.0	12.9	-17.1
Tennessee . . .	49	54	58	45	56	60	54	69	73	70	66	59	51	63	64	30.6	22.4	10.0	-3.0
Texas	140	119	139	147	166	198	172	123	154	169	129	175	189	164	203	45.0	41.4	-34.8	57.4
Virginia	56	69	69	60	72	93	77	72	80	98	103	84	96	101	109	94.6	66.1	10.8	5.8
West Virginia .	28	27	27	17	27	20	27	5	8	6	5	3	7	3	3	-89.3	-28.6	-75.0	-40.0
West	728	755	723	730	821	845	815	820	802	767	834	776	799	795	786	8.0	16.1	-1.3	-5.8
Alaska	2	1	0	2	0	1	1	0	1	3	0	—	1	3	2	0.0	-50.0	-100.0	(1)
Arizona	51	59	52	40	63	34	49	46	51	42	42	62	61	58	54	5.9	-33.3	23.5	28.6
California	436	436	431	413	467	515	501	479	484	468	501	444	482	454	478	9.6	18.1	-2.7	-4.6
Colorado	47	53	49	53	61	47	52	70	47	68	69	66	64	69	50	6.4	0.0	46.8	-27.5
Hawaii	16	19	28	18	18	30	25	23	18	16	21	16	22	25	14	-12.5	87.5	-30.0	-33.3
Idaho	10	7	9	13	11	9	11	9	8	9	4	7	9	6	14	40.0	-10.0	-55.6	250.0

Table 5-23.—Continued

Region and state	1975-76	1976-77	1977-78	1978-79	1979-80	1980-81	1981-82	1982-83	1983-84	1984-85	1985-86	1986-87	1987-88	1988-89	1989-90	Percent change 1975-76 to 1989-90	Percent change 1975-76 to 1980-81	Percent change 1980-81 to 1985-86	Percent change 1985-86 to 1989-90
West Continued																			
Montana	15	11	7	7	7	5	6	13	9	7	18	9	8	5	7	-53.3	-66.7	260.0	-61.1
Nevada	3	3	2	3	5	5	3	3	0	3	5	2	6	4	7	133.3	66.7	0.0	40.0
New Mexico	7	7	8	7	13	12	9	7	11	9	16	11	9	12	9	28.6	71.4	33.3	-43.8
Oregon	34	40	39	47	36	69	36	51	51	40	47	39	24	38	46	35.3	102.9	-31.9	-2.1
Utah	36	44	28	40	48	37	44	38	45	24	26	27	27	30	23	-36.1	2.8	-29.7	-11.5
Washington	66	66	66	80	79	74	71	71	75	74	72	81	79	83	77	16.7	12.1	-2.7	6.9
Wyoming	5	9	4	7	13	7	7	10	2	4	13	12	7	8	5	0.0	40.0	85.7	-61.5
U.S. Service Schools	0	0	0	0	0	0	2	5	5	13	8	8	—	—	—	([1])	([1])	([1])	([1])
Outlying Areas	5	6	4	6	2	6	5	4	2	4	7	1	11	5	7	40.0	20.0	16.7	0.0
American Samoa	—	—	—	—	—	—	—	—	—	—	—	—	—	—	—	([1])	([1])	([1])	([1])
Guam	0	0	0	0	0	0	0	0	0	0	0	—	—	—	—	([1])	([1])	([1])	([1])
Northern Marianas	—	—	—	—	—	—	—	—	—	—	—	—	—	—	—	([1])	([1])	([1])	([1])
Puerto Rico	5	6	4	6	2	6	5	4	2	4	7	1	11	5	7	40.0	20.0	16.7	0.0
Trust Territories	—	—	—	—	—	—	—	—	—	—	—	—	—	—	—	([1])	([1])	([1])	([1])
Virgin Islands	0	0	0	0	0	0	0	0	0	0	0	—	—	—	—	([1])	([1])	([1])	([1])

[1] Insufficient data for calculating a percent change.

—Data not available.

SOURCE: U.S. Department of Education, National Center for Education Statistics, HEGIS, "Degrees and Other Formal Awards Conferred" surveys; and IPEDS, "Completions" surveys.

Table 5-24.—Doctor's degrees conferred in physical sciences, by region and state: 1975-76 to 1989-90

Region and state	1975-76	1976-77	1977-78	1978-79	1979-80	1980-81	1981-82	1982-83	1983-84	1984-85	1985-86	1986-87	1987-88	1988-89	1989-90	Percent change 1975-76 to 1989-90	Percent change 1975-76 to 1980-81	Percent change 1980-81 to 1985-86	Percent change 1985-86 to 1989-90
United States . . .	3,431	3,341	3,133	3,102	3,089	3,141	3,286	3,269	3,306	3,403	3,551	3,673	3,809	3,858	4,168	21.5	-8.5	13.1	17.4
Northeast	1,025	964	895	918	882	929	927	929	994	956	1,014	1,013	986	1,078	1,155	12.7	-9.4	9.1	13.9
Connecticut . .	52	47	50	49	59	39	42	63	69	44	70	66	63	67	85	63.5	-25.0	79.5	21.4
Maine	2	4	3	1	4	4	5	3	2	2	5	6	7	4	8	300.0	100.0	25.0	60.0
Massachusetts	286	257	216	247	216	268	236	243	236	263	288	227	265	259	305	6.6	-6.3	7.5	5.9
New Hampshire	19	20	13	11	23	15	17	16	14	17	14	22	17	16	21	10.5	-21.1	-6.7	50.0
New Jersey . . .	76	92	85	89	74	88	88	91	100	95	92	86	100	97	102	34.2	15.8	4.5	10.9
New York	352	318	297	315	296	304	319	303	314	301	316	334	304	371	398	13.1	-13.6	3.9	25.9
Pennsylvania .	188	175	177	152	154	167	172	156	195	176	171	211	162	197	184	-2.1	-11.2	2.4	7.6
Rhode Island .	40	48	47	48	49	39	43	51	59	51	53	53	58	60	47	17.5	-2.5	35.9	-11.3
Vermont	10	3	7	6	7	5	5	3	5	7	5	8	10	7	5	-50.0	-50.0	0.0	0.0
Midwest	889	836	792	814	749	779	833	810	816	827	866	942	958	958	989	11.2	-12.4	11.2	14.2
Illinois	202	202	196	217	189	220	196	199	176	204	199	225	201	209	236	16.8	8.9	-9.5	18.6
Indiana	112	122	87	95	95	99	90	110	112	110	97	116	114	127	136	21.4	-11.6	-2.0	40.2
Iowa	70	56	50	59	53	53	65	52	56	59	60	80	71	60	57	-18.6	-24.3	13.2	-5.0
Kansas	32	27	45	21	25	28	31	33	31	34	25	34	29	27	36	12.5	-12.5	-10.7	44.0
Michigan	106	92	113	111	89	77	102	106	116	95	122	126	129	140	124	17.0	-27.4	58.4	1.6
Minnesota	33	32	40	25	26	33	34	29	48	38	53	40	48	43	49	48.5	0.0	60.6	-7.5
Missouri	44	62	36	30	43	39	55	24	44	43	41	41	46	45	53	20.5	-11.4	5.1	29.3
Nebraska	12	13	15	20	19	22	19	29	15	17	17	20	25	28	18	50.0	83.3	-22.7	5.9
North Dakota .	13	9	6	11	9	9	10	7	7	4	14	13	14	24	12	-7.7	-30.8	55.6	-14.3
Ohio	162	131	132	123	123	130	142	126	134	141	140	147	177	153	163	0.6	-19.8	7.7	16.4
South Dakota .	2	0	1	0	1	1	2	0	0	0	1	1	2	1	2	0.0	-50.0	0.0	100.0
Wisconsin	101	90	71	102	77	68	87	95	77	82	97	99	102	101	103	2.0	-32.7	42.6	6.2
South	730	733	669	682	653	602	673	663	643	732	757	802	832	839	974	33.4	-17.5	25.7	28.7
Alabama	15	14	18	11	8	15	11	17	15	10	14	20	11	30	30	100.0	0.0	-6.7	114.3
Arkansas	9	7	5	9	8	9	12	12	9	9	12	16	16	6	14	55.6	0.0	33.3	16.7
Delaware	15	16	17	9	17	11	32	10	0	19	21	16	16	27	27	80.0	-26.7	90.9	28.6
District of Columbia . . .	43	45	47	45	36	33	38	40	35	32	37	31	32	31	31	-27.9	-23.3	12.1	-16.2
Florida	74	84	64	76	62	51	47	58	47	70	53	67	82	89	97	31.1	-31.1	3.9	83.0
Georgia	38	35	34	39	40	38	59	34	40	54	48	56	42	54	53	39.5	0.0	26.3	10.4
Kentucky	11	7	15	5	11	8	4	10	6	10	10	11	17	9	21	90.9	-27.3	25.0	110.0
Louisiana	25	29	24	16	22	19	21	28	26	30	32	34	32	36	53	112.0	-24.0	68.4	65.6
Maryland	84	84	79	85	57	68	64	73	58	78	70	81	74	62	85	1.2	-19.0	2.9	21.4
Mississippi . . .	11	16	12	2	11	2	15	11	14	10	14	15	13	13	21	90.9	-81.8	600.0	50.0
North Carolina	66	72	61	62	66	61	66	67	81	76	83	84	92	49	91	37.9	-7.6	36.1	9.6
Oklahoma	31	32	20	25	28	21	26	19	19	20	33	28	25	22	36	16.1	-32.3	57.1	9.1
South Carolina	29	33	31	38	36	28	27	29	31	39	38	44	57	44	52	79.3	-3.4	35.7	36.8
Tennessee . . .	27	35	32	32	29	27	16	27	25	23	24	27	37	24	28	3.7	0.0	-11.1	16.7
Texas	195	180	164	181	175	162	176	189	178	199	204	214	202	250	253	29.7	-16.9	25.9	24.0
Virginia	52	37	37	40	34	44	49	33	54	42	58	56	75	81	74	42.3	-15.4	31.8	27.6
West Virginia .	5	7	9	7	13	5	10	6	5	11	6	2	9	12	8	60.0	0.0	20.0	33.3
West	785	806	777	688	805	830	850	865	851	885	912	916	1,033	983	1,050	33.8	5.7	9.9	15.1
Alaska	4	2	2	3	0	1	0	3	4	5	6	3	10	6	3	-25.0	-75.0	500.0	-50.0
Arizona	57	56	56	52	56	61	77	72	64	76	72	63	94	106	93	63.2	7.0	18.0	29.2
California	436	425	432	383	457	491	490	496	494	504	528	545	550	511	554	27.1	12.6	7.5	4.9
Colorado	73	69	68	69	76	82	70	82	85	87	91	81	103	101	125	71.2	12.3	11.0	37.4
Hawaii	11	20	25	16	17	14	16	13	13	19	16	22	18	32	14	27.3	27.3	14.3	-12.5
Idaho	15	12	5	2	6	4	6	11	8	3	5	6	7	10	8	-46.7	-73.3	25.0	60.0

Table 5-24.—Continued

Region and state	1975-76	1976-77	1977-78	1978-79	1979-80	1980-81	1981-82	1982-83	1983-84	1984-85	1985-86	1986-87	1987-88	1988-89	1989-90	Percent change 1975-76 to 1989-90	1975-76 to 1980-81	1980-81 to 1985-86	1985-86 to 1989-90
West Continued																			
Montana	8	8	5	6	5	2	8	12	10	12	8	7	13	14	18	125.0	-75.0	300.0	125.0
Nevada	5	13	8	4	6	2	7	5	6	1	2	3	4	6	10	100.0	-60.0	0.0	400.0
New Mexico ..	26	22	31	26	22	27	17	26	24	29	25	29	34	44	32	23.1	3.8	-7.4	28.0
Oregon	50	68	47	40	47	31	33	36	49	39	41	38	49	31	48	-4.0	-38.0	32.3	17.1
Utah.........	28	39	39	24	38	35	37	35	26	34	35	51	67	37	49	75.0	25.0	0.0	40.0
Washington...	61	57	47	57	64	68	72	61	55	64	66	61	66	73	77	26.2	11.5	-2.9	16.7
Wyoming.....	11	15	12	6	11	12	17	13	13	12	17	7	18	12	19	72.7	9.1	41.7	11.8
U.S. Service Schools......	2	2	0	0	0	1	3	2	2	3	2	0	0	0	0	-100.0	-50.0	100.0	-100.0
Outlying Areas..	2	3	4	2	6	4	4	1	4	4	2	6	7	6	6	200.0	100.0	-50.0	200.0
American Samoa.....	—	—	—	—	—	—	—	—	—	—	—	—	—	—	—	(1)	(1)	(1)	(1)
Guam	0	0	0	—	—	0	0	0	0	—	0	—	—	—	—	(1)	(1)	(1)	(1)
Northern Marianas...	—	—	—	—	—	—	—	—	—	—	—	—	—	—	—	(1)	(1)	(1)	(1)
Puerto Rico ..	2	3	4	2	6	4	4	1	4	4	2	6	7	6	6	200.0	100.0	-50.0	200.0
Trust Territories	—	—	—	—	—	—	—	—	—	—	—	—	—	—	—	(1)	(1)	(1)	(1)
Virgin Islands .	—	0	0	0	0	0	0	0	0	0	0	—	—	—	—	(1)	(1)	(1)	(1)

[1] Insufficient data for calculating a percent change.

—Data not available.

SOURCE: U.S. Department of Education, National Center for Education Statistics, HEGIS, "Degrees and Other Formal Awards Conferred" surveys; and IPEDS, "Completions" surveys.

THIS PAGE INTENTIONALLY LEFT BLANK

6. Supply of Graduates in Science and Mathematics: Nonresident Aliens Receiving Degrees in Science and Mathematics

A foreign student, or "nonresident alien," is a person who is not a citizen or national of the United States and who is in this country on a visa or temporary basis and does not have the right to remain indefinitely. Nonresident aliens studying in U.S. colleges increased steadily throughout the 1980s, rising from 305,000 in 1980–81 to 397,000 in 1989–90.[1] The number of nonresident aliens receiving any type of degree in science or mathematics rose from 22,000 in 1980–81 to approximately 34,000 in 1989–90.[2] The proportions of the science and mathematics degrees conferred in the U.S. to nonresident aliens were small at the associate and bachelor's degree levels. They were more notable at the master's and doctor's degree levels, and were also more notable in science and mathematics than in other fields.

Associate Degrees

Table 6–1 shows that the proportions of associate degrees in science[3] going to nonresident aliens in 1989–90 were low, ranging from less than one percent in the health sciences to over 2 percent in life sciences. There was little change from 1984–85 when the proportion of degrees was less than 1 percent in health sciences and 1 or 2 percent in all other science fields. The percentages of nonresident aliens receiving associate degrees in mathematics dropped from almost 6 percent in 1984–85 to 2 percent in 1989–90. The percentages of associate degrees going to nonresident aliens in other fields such as business and management, education, psychology, public affairs and services, and social sciences were as small as the percentage of associate degrees awarded to nonresident aliens in the science fields. The percentage of associates in science and mathematics going to nonresident aliens was about the same as the percentage in fields other than science and mathematics.

[1] National Center for Education Statistics. *Digest of Education Statistics, 1992.* (Washington, D.C.: U.S. Government Printing Office, 1992) p. 203.

[2] National Center for Education Statistics. *Race/Ethnicity Trends in Degrees Conferred by Institutions of Higher Education: 1980–81 through 1989–90.* (Washington, D.C.: U.S. Government Printing Office, 1992) pp. 14, 18, 22, 26.

[3] In this chapter, "science" refers to the following fields: agricultural sciences, computer sciences, engineering, health sciences, life sciences, and physical sciences, for all tables except table 6–5 (see footnote 4).

Bachelor's Degrees

For the years 1980–81 to 1989–90, about 2 to 3 percent of all bachelor's degrees, compared to 4 to 5 percent of bachelor's degrees in science and mathematics, were conferred upon nonresident aliens (table 6–2). About 7 percent of all bachelor's degrees in engineering in 1989–90 went to nonresident aliens, down from over 9 percent in 1980–81. The proportion of bachelor's degrees to nonresident aliens in computer sciences went up from 5 percent in 1980–81 to 8 percent in 1989–90. In all other science fields the percentages of bachelor's degrees awarded to nonresident aliens in 1989–90 were lower, ranging from 1 percent in health sciences to 4 percent in physical sciences. The percentage of bachelor's in mathematics going to nonresident aliens was 4 percent in both 1980–81 and 1989–90.

The percentage of bachelor's degrees in 1989–90 received by nonresident aliens in the fields of education, psychology, public affairs and services, and social sciences ranged from 1 percent in education to 2 percent in social sciences. In business and management, the percentage of nonresident aliens receiving bachelor's degrees was 3 percent.

Master's Degrees

A larger proportion of graduate degrees was conferred upon nonresident aliens in 1989–90 than associate and bachelor's degrees, particularly in science fields and mathematics (tables 6–3 and 6–4). In 1989–90, nonresident aliens received 21 percent of all master's degrees conferred in science and mathematics, up from 16 percent in 1980–81. During that same year over 26 percent of all master's degrees conferred in agricultural sciences went to nonresident aliens; this was up from almost 18 percent in 1980–81. Nonresident aliens received approximately 26 percent of the master's degrees in physical sciences, up from 15 percent in 1980–81. The proportion of master's degrees awarded to nonresident aliens was even higher in computer sciences at 29 percent, up from 22 percent in 1980–81. The proportion receiving mathematics degrees was 29 percent, up from 18 percent in 1980–81 and the proportion in engineering was 30 percent, up slightly from 28 percent in 1980–81. Health sciences and life sciences awarded smaller proportions of master's degrees to nonresident aliens in 1989–90, 5 percent and 15 percent, respectively.

The proportions of master's degrees awarded to nonresident aliens during 1989–90 in other fields were smaller than in the science and mathematics fields, with the exception of the social sciences (20 percent). In 1989–90 nonresident aliens received 3 percent of master's in education, 4 percent in psychology and public affairs and services, and 10 percent of all master's degrees in business and management.

Doctor's Degrees

The large proportion of doctor's degrees in science and mathematics awarded to nonresident aliens has caused some concern among scientists, educators, and policymakers following science education statistics. Some areas of concern include whether these students are expanding or draining U.S. educational resources, whether these students are replacing potential U.S. students, and whether in educating these students, the U.S. may be aiding future economic competitors in the global marketplace.

About 35 percent of all doctor's degrees in science and mathematics were conferred upon nonresident aliens in 1989–90. This was up from 20 percent in 1980–81. Some fields were much higher. In engineering, 50 percent of doctor's degrees were awarded to nonresident aliens in 1989–90, up from 37 percent in 1980–81. For mathematics, the rise in the proportion of doctor's degrees was even more dramatic, from 24 percent in 1980–81 to 51 percent in 1989–90.

In the field of computer sciences, over 44 percent of all doctor's degrees were awarded to nonresident aliens, up from 21 percent in 1980–81. In 1989–90, 39 percent of all doctor's degrees in agricultural sciences were conferred to nonresident aliens and almost a third of all doctor's degrees in physical sciences were awarded to nonresident aliens that same year. Similar to the master's level, health sciences and life sciences awarded the lowest proportions (about 20 percent) of doctor's degrees in the science fields to nonresident aliens in 1989–90.

Nonresident aliens account for a much smaller proportion of doctor's degrees in fields other than science and mathematics. However, some non-science and mathematics fields also had large proportions of nonresident aliens receiving doctor's degrees. Over a third of all doctor's degrees in business and management were received by nonresident aliens in 1989–90, and one quarter of all doctor's degrees in social sciences were awarded to nonresident aliens. In other fields the percentages of 1989–90 doctor's degrees awarded to nonresident aliens were much smaller: 15 percent for public affairs and services, almost 9 percent for education, and 4 percent in psychology.

Postgraduation Plans for Doctor's Degree Recipients

Data from the National Science Foundation (NSF) yearly Survey of Earned Doctorates (SED) show that 33 percent of nonresident aliens receiving doctor's degrees in 1991 said they planned to stay in the United States to pursue postdoctoral study, to work in academia, or to work in industry or some other area (table 6–5). This percentage is up from 25 percent in 1980. The percentage of science and engineering doctor's degree recipients[4] staying in the U.S. after graduation was 35 percent in 1991, up from 30 percent in 1980. For non-science and mathematics fields, 25 percent planned to stay in the U.S. after graduation in 1991, up from 10 percent in 1980. Of those 1991 science and engineering doctor's degree recipients planning to stay in the U.S., 53 percent planned to pursue postdoctoral study, 21 percent planned to work in academia, 22 percent planned to work in industry, and 4 percent planned for other employment.

Table 6–5 represents the immediate plans of nonresident doctor's degree recipients. In the long term, more of these degree recipients could return to their home country, and some who left immediately could return to the U.S.

Summary

Tables 6–3 and 6–4 document the increasing proportions of graduate degrees going to nonresident aliens throughout the 1980s. The increases were generally greater in the science fields and mathematics than in the other fields. According to the 1991 edition of *The Condition of Education*: "Growth in the foreign student population can affect enrollment levels and in turn, influence the amount and allocation of material, personnel, and financial resources. It may also signal potential problems for U.S. economic competitiveness, depending on changes in the number of Americans receiving degrees in critical fields and on whether the foreign students stay in this country to work after completing their studies...The decline in American doctorate recipients occurred during a period [1977–1989] of growth (36 percent) in the 25- to 34-year old college graduate population. Although the number of American recipients has increased since 1985, growth has been slower than growth in the number of college graduates aged 25–34."[5]

The decline in the numbers of American students pursuing graduate degrees in science fields, and the growth in the numbers of nonresident aliens studying in science fields suggest the need for further research.

[4] The category "science and engineering" in this survey included social sciences and psychology.

[5] National Center for Education Statistics. *The Condition of Education, 1991*. (Washington, D.C.: U.S. Government Printing Office, 1991) Volume 2, p. 68.

THIS PAGE INTENTIONALLY LEFT BLANK

Table 6-1.—Nonresident aliens receiving associate degrees, by field: 1984-85 to 1989-90

Field of study	1980-81	1984-85	1986-87	1988-89	1989-90	Percent change 1984-85 to 1989-90
All associate degrees						
All students	—	429,185	436,299	432,144	448,997	4.6
U.S. students	—	422,778	431,611	425,782	442,820	4.7
Nonresident aliens	—	6,407	4,688	6,362	6,177	-3.6
Nonresident aliens as a percent of all students	—	1.49	1.07	1.47	1.38	[1]-0.12
Agricultural sciences						
All students	—	6,320	5,458	4,725	4,832	-23.5
U.S. students	—	6,243	5,414	4,666	4,734	-24.2
Nonresident aliens	—	77	44	59	98	27.3
Nonresident aliens as a percent of all students	—	1.22	0.81	1.25	2.03	[1]0.81
Computer sciences						
All students	—	11,843	9,101	7,900	7,604	-35.8
U.S. students	—	11,570	8,930	7,698	7,444	-35.7
Nonresident aliens	—	273	171	202	160	-41.4
Nonresident aliens as a percent of all students	—	2.31	1.88	2.56	2.10	[1]-0.20
Engineering						
All students	—	59,391	62,512	56,368	54,131	-8.9
U.S. students	—	58,417	61,737	55,630	53,490	-8.4
Nonresident aliens	—	974	775	738	641	-34.2
Nonresident aliens as a percent of all students	—	1.64	1.24	1.31	1.18	[1]-0.46
Health sciences						
All students	—	65,864	62,547	59,566	64,128	-2.6
U.S. students	—	65,488	62,231	59,136	63,678	-2.8
Nonresident aliens	—	376	316	430	450	19.7
Nonresident aliens as a percent of all students	—	0.57	0.51	0.72	0.70	[1]0.13
Life sciences						
All students	—	852	892	982	1,034	21.4
U.S. students	—	832	856	946	1,009	21.3
Nonresident aliens	—	20	36	36	25	25.0
Nonresident aliens as a percent of all students	—	2.35	4.04	3.67	2.42	[1]0.07
Physical sciences						
All students	—	1,999	2,061	1,961	2,135	6.8
U.S. students	—	1,970	2,022	1,912	2,098	6.5
Nonresident aliens	—	29	39	49	37	27.6
Nonresident aliens as a percent of all students	—	1.45	1.89	2.50	1.73	[1]0.28
Total - associate degrees in science						
All students	—	154,830	142,571	131,502	133,864	-13.5
U.S. students	—	153,081	141,190	129,988	132,453	-13.5
Nonresident aliens	—	1,749	1,381	1,514	1,411	-19.3
Nonresident aliens as a percent of all students	—	1.13	0.97	1.15	1.05	[1]-0.08
Mathematics						
All students	—	693	666	654	760	9.7
U.S. students	—	654	654	636	742	13.5
Nonresident aliens	—	39	12	18	18	-53.8
Nonresident aliens as a percent of all students	—	5.63	1.80	2.75	2.37	[1]-3.26
Total - associate degrees in science and mathematics						
All students	—	155,619	143,238	132,156	134,624	-13.5
U.S. students	—	153,831	141,845	130,624	133,195	-13.4
Nonresident aliens	—	1,788	1,393	1,532	1,429	-20.1
Nonresident aliens as a percent of all students	—	1.15	0.97	1.16	1.06	[1]-0.09
Non-Science and Mathematics Fields						
Business and management						
All students	—	116,737	115,231	107,629	106,980	-8.4
U.S. students	—	114,961	114,047	106,058	105,484	-8.2
Nonresident aliens	—	1,776	1,184	1,571	1,496	-15.8
Nonresident aliens as a percent of all students	—	1.52	1.03	1.46	1.40	[1]-0.12
Education						
All students	—	7,009	7,333	7,391	8,018	14.4
U.S. students	—	6,948	7,259	7,285	7,945	14.3
Nonresident aliens	—	61	74	106	73	19.7
Nonresident aliens as a percent of all students	—	0.87	1.01	1.43	0.91	[1]0.04

Table 6-1.—Continued

Field of study	1980-81	1984-85	1986-87	1988-89	1989-90	Percent change 1984-85 to 1989-90
Psychology						
All students	—	821	1,014	1,090	1,110	35.2
U.S. students	—	807	1,006	1,065	1,103	36.7
Nonresident aliens	—	14	8	25	7	-50.0
Nonresident aliens as a percent of all students	—	1.71	0.79	2.29	0.63	[1]-1.07
Public affairs and services						
All students	—	16,105	16,021	16,875	18,537	15.1
U.S. students	—	15,993	15,925	16,770	18,427	15.2
Nonresident aliens	—	112	96	105	110	-1.8
Nonresident aliens as a percent of all students	—	0.70	0.60	0.62	0.59	[1]-0.10
Social sciences						
All students	—	2,086	2,584	2,741	2,870	37.6
U.S. students	—	2,027	2,560	2,688	2,836	39.9
Nonresident aliens	—	59	24	53	34	-42.4
Nonresident aliens as a percent of all students	—	2.83	0.93	1.93	1.18	[1]-1.64
Other						
All students	—	130,808	150,878	164,262	176,858	35.2
U.S. students	—	128,211	148,969	161,292	173,830	35.6
Nonresident aliens	—	2,597	1,909	2,970	3,028	16.6
Nonresident aliens as a percent of all students	—	1.99	1.27	1.81	1.71	[1]-0.27

[1] This figure represents the change in percentage points from 1984-85 to 1989-90.

— Comparable data on associate degrees by field and ethnicity before 1984-85 were not reported.

NOTE—Data for 1984-85 do not include imputations for nonresident aliens. Data for 1988-89 have been revised from previously published figures. Data represent programs, not organizational units within institutions. Because of adjustments to underreported and nonreported data, figures for degrees for all students for 1984-85 and 1986-87 in this table are slightly different from corresponding data in other tables. Data for Public affairs and services are different from data in the *Digest of Education Statistics* due to the use of different subcategories.

SOURCE: U.S. Department of Education, National Center for Education Statistics, HEGIS, "Degrees and Other Formal Awards Conferred" surveys; and IPEDS, "Completions" and "Consolidated" surveys.

Table 6-2.—Nonresident aliens receiving bachelor's degrees, by field: 1980-81 to 1989-90

Field of study	1980-81	1984-85	1986-87	1988-89	1989-90	Percent change 1980-81 to 1989-90
All bachelor's degrees						
All students	934,800	968,311	991,260	1,016,350	1,046,930	12.0
U.S. students	912,211	939,094	961,954	989,314	1,020,153	11.8
Nonresident aliens	22,589	29,217	29,306	27,036	26,777	18.5
Nonresident aliens as a percent of all students	2.42	3.02	2.96	2.66	2.56	[1]0.14
Agricultural sciences						
All students	21,886	18,015	14,991	13,492	13,070	-40.3
U.S. students	21,270	17,356	14,528	13,094	12,691	-40.3
Nonresident aliens	616	659	463	398	379	-38.5
Nonresident aliens as a percent of all students	2.81	3.66	3.09	2.95	2.90	[1]0.09
Computer sciences						
All students	15,120	38,589	39,590	30,454	27,434	81.4
U.S. students	14,343	36,473	36,918	28,244	25,321	76.5
Nonresident aliens	777	2,116	2,672	2,210	2,113	171.9
Nonresident aliens as a percent of all students	5.14	5.48	6.75	7.26	7.70	[1]2.56
Engineering						
All students	74,954	94,560	93,097	85,225	82,110	9.5
U.S. students	67,991	87,165	86,128	79,410	76,465	12.5
Nonresident aliens	6,963	7,395	6,969	5,815	5,645	-18.9
Nonresident aliens as a percent of all students	9.29	7.82	7.49	6.82	6.87	[1]-2.41
Health sciences						
All students	63,649	63,289	63,213	59,138	58,816	-7.6
U.S. students	63,067	62,470	62,415	58,380	58,012	-8.0
Nonresident aliens	582	819	798	758	804	38.1
Nonresident aliens as a percent of all students	0.91	1.29	1.26	1.28	1.37	[1]0.45
Life sciences						
All students	43,216	38,115	38,120	36,059	37,170	-14.0
U.S. students	42,315	37,204	37,237	35,174	36,299	-14.2
Nonresident aliens	901	911	883	885	871	-3.3
Nonresident aliens as a percent of all students	2.08	2.39	2.32	2.45	2.34	[1]0.26
Physical sciences						
All students	23,950	23,555	20,071	17,186	16,131	-32.6
U.S. students	23,218	22,767	19,418	16,575	15,530	-33.1
Nonresident aliens	732	788	653	611	601	-17.9
Nonresident aliens as a percent of all students	3.06	3.35	3.25	3.56	3.73	[1]0.67
Total - bachelor's degrees in science						
All students	242,253	279,780	269,080	241,554	234,731	-3.1
U.S. students	231,682	267,092	256,642	230,877	224,318	-3.2
Nonresident aliens	10,571	12,688	12,438	10,677	10,413	-1.5
Nonresident aliens as a percent of all students	4.36	4.53	4.62	4.42	4.44	[1]0.07
Mathematics						
All students	11,078	14,885	16,444	15,218	14,597	31.8
U.S. students	10,623	14,124	15,774	14,672	14,073	32.5
Nonresident aliens	455	761	670	546	524	15.2
Nonresident aliens as a percent of all students	4.11	5.02	4.07	3.61	3.59	[1]-0.52
Total - bachelor's degrees in science and mathematics						
All students	253,331	294,926	285,524	256,682	249,328	-1.6
U.S. students	242,305	281,477	272,416	245,459	238,391	-1.6
Nonresident aliens	11,026	13,449	13,108	11,223	10,937	-0.8
Nonresident aliens as a percent of all students	4.35	4.56	4.59	4.37	4.39	[1]0.03
Non-Science and Mathematics Fields						
Business and management						
All students	200,857	231,308	241,100	247,175	249,081	24.0
U.S. students	196,291	223,880	232,986	239,217	241,460	23.0
Nonresident aliens	4,566	7,428	8,114	7,958	7,621	66.9
Nonresident aliens as a percent of all students	2.27	3.21	3.37	3.22	3.06	[1]34.6
Education						
All students	108,265	87,788	87,083	97,082	104,715	-3.3
U.S. students	107,357	86,773	86,236	96,441	104,038	-3.1
Nonresident aliens	908	1,015	847	641	677	-25.4
Nonresident aliens as a percent of all students	0.84	1.16	0.97	0.66	0.65	[1]-22.9

Table 6-2.—Continued

Field of study	1980-81	1984-85	1986-87	1988-89	1989-90	Percent change 1980-81 to 1989-90
Psychology						
All students	40,833	39,523	42,835	48,737	53,586	31.2
U.S. students	40,349	38,980	42,324	48,210	53,053	31.5
Nonresident aliens	484	543	511	527	533	10.1
Nonresident aliens as a percent of all students	1.19	1.37	1.19	1.08	0.99	[1]-16.1
Public affairs and services						
All students	36,311	30,572	31,214	34,151	36,032	-0.8
U.S. students	35,995	30,159	30,882	33,845	35,710	-0.8
Nonresident aliens	316	413	332	306	322	1.9
Nonresident aliens as a percent of all students	0.87	1.35	1.06	0.90	0.89	[1]2.7
Social sciences						
All students	100,647	90,795	96,172	107,914	116,925	16.2
U.S. students	98,671	88,544	93,890	105,678	114,475	16.0
Nonresident aliens	1,976	2,251	2,282	2,236	2,450	24.0
Nonresident aliens as a percent of all students	1.96	2.48	2.37	2.07	2.10	[1]6.7
Other						
All students	194,556	193,399	207,332	224,609	237,263	22.0
U.S. students	191,243	189,281	203,220	220,464	233,026	21.8
Nonresident aliens	3,313	4,118	4,112	4,145	4,237	27.9
Nonresident aliens as a percent of all students	1.70	2.13	1.98	1.85	1.79	[1]4.9

[1] This figure represents the change in percentage points from 1980-81 to 1989-90.

NOTE—Data for 1980-81 and 1984-85 do not include imputations for nonresident aliens. Data for 1988-89 have been revised from previously published figures. Data represent programs, not organizational units within institutions. Because of adjustments to underreported and nonreported data, figures for degrees for all students for 1980-81, 1984-85 and 1986-87 in this table are slightly different from corresponding data in other tables. Data for Public affairs and services are different from data in the *Digest of Education Statistics* due to the use of different subcategories.

SOURCE: U.S. Department of Education, National Center for Education Statistics, HEGIS, "Degrees and Other Formal Awards Conferred" surveys; and IPEDS, "Completions" and "Consolidated" surveys.

Table 6-3.—Nonresident aliens receiving master's degrees, by field: 1980-81 to 1989-90

Field of study	1980-81	1984-85	1986-87	1988-89	1989-90	Percent change 1980-81 to 1989-90
All master's degrees						
All students	294,183	280,421	289,341	309,770	321,992	9.5
U.S. students	272,126	253,469	259,443	275,556	286,508	5.3
Nonresident aliens	22,057	26,952	29,898	34,214	35,484	60.9
Nonresident aliens as a percent of all students	7.50	9.61	10.33	11.04	11.02	[1]3.52
Agricultural sciences						
All students	4,003	3,901	3,521	3,245	3,373	-15.7
U.S. students	3,293	3,085	2,723	2,389	2,484	-24.6
Nonresident aliens	710	816	798	856	889	25.2
Nonresident aliens as a percent of all students	17.74	20.92	22.66	26.38	26.36	[1]8.62
Computer sciences						
All students	4,143	6,942	8,481	9,414	9,643	132.8
U.S. students	3,239	5,233	6,264	6,661	6,889	112.7
Nonresident aliens	904	1,709	2,217	2,753	2,754	204.6
Nonresident aliens as a percent of all students	21.82	24.62	26.14	29.24	28.56	[1]6.74
Engineering						
All students	16,358	20,735	22,658	24,572	24,848	51.9
U.S. students	11,795	14,922	16,547	17,259	17,287	46.6
Nonresident aliens	4,563	5,813	6,111	7,313	7,561	65.7
Nonresident aliens as a percent of all students	27.89	28.03	26.97	29.76	30.43	[1]2.53
Health sciences						
All students	16,515	17,062	18,421	19,293	20,354	23.2
U.S. students	15,817	16,217	17,509	18,164	19,248	21.7
Nonresident aliens	698	845	912	1,129	1,106	58.5
Nonresident aliens as a percent of all students	4.23	4.95	4.95	5.85	5.43	[1]1.21
Life sciences						
All students	5,978	5,010	4,950	4,961	4,861	-18.7
U.S. students	5,610	4,536	4,414	4,288	4,127	-26.4
Nonresident aliens	368	474	536	673	734	99.5
Nonresident aliens as a percent of all students	6.16	9.46	10.83	13.57	15.10	[1]8.94
Physical sciences						
All students	5,227	5,675	5,630	5,723	5,447	4.2
U.S. students	4,441	4,575	4,512	4,406	4,041	-9.0
Nonresident aliens	786	1,100	1,118	1,317	1,406	78.9
Nonresident aliens as a percent of all students	15.04	19.38	19.86	23.01	25.81	[1]10.78
Total - master's degrees in science						
All students	52,196	60,824	63,662	67,208	68,526	31.3
U.S. students	44,167	50,067	51,970	53,167	54,076	22.4
Nonresident aliens	8,029	10,757	11,692	14,041	14,450	80.0
Nonresident aliens as a percent of all students	15.38	17.69	18.37	20.89	21.09	[1]5.70
Mathematics						
All students	2,565	2,831	3,319	3,447	3,677	43.4
U.S. students	2,101	2,146	2,440	2,416	2,616	24.5
Nonresident aliens	464	685	879	1,031	1,061	128.7
Nonresident aliens as a percent of all students	18.09	24.20	26.48	29.91	28.86	[1]10.77
Total - master's degrees in science and mathematics						
All students	54,761	63,655	66,981	70,655	72,203	31.9
U.S. students	46,268	52,213	54,410	55,583	56,692	22.5
Nonresident aliens	8,493	11,442	12,571	15,072	15,511	82.6
Nonresident aliens as a percent of all students	15.51	17.98	18.77	21.33	21.48	[1]5.97
Non-Science and Mathematics Fields						
Business and management						
All students	57,541	66,596	67,504	73,521	77,203	34.2
U.S. students	52,490	60,780	60,303	65,533	69,167	31.8
Nonresident aliens	5,051	5,816	7,201	7,988	8,036	59.1
Nonresident aliens as a percent of all students	8.78	8.73	10.67	10.86	10.41	[1]1.63
Education						
All students	98,380	75,821	75,473	82,533	86,057	-12.5
U.S. students	95,681	72,902	73,074	79,965	83,092	-13.2
Nonresident aliens	2,699	2,919	2,399	2,568	2,965	9.9
Nonresident aliens as a percent of all students	2.74	3.85	3.18	3.11	3.45	[1]0.70

Table 6-3.—Continued

Field of study	1980-81	1984-85	1986-87	1988-89	1989-90	Percent change 1980-81 to 1989-90
Psychology						
All students	7,998	8,379	8,124	8,552	9,231	15.4
U.S. students	7,728	8,083	7,891	8,274	8,841	14.4
Nonresident aliens	270	296	233	278	390	44.4
Nonresident aliens as a percent of all students	3.38	3.53	2.87	3.25	4.22	[1]0.85
Public affairs and services						
All students	20,074	17,130	18,523	19,417	19,574	-2.5
U.S. students	19,355	16,426	17,694	18,467	18,735	-3.2
Nonresident aliens	719	704	829	950	839	16.7
Nonresident aliens as a percent of all students	3.58	4.11	4.48	4.89	4.29	[1]0.70
Social sciences						
All students	11,917	10,223	10,395	10,867	11,419	-4.2
U.S. students	10,322	8,398	8,375	8,713	9,162	-11.2
Nonresident aliens	1,595	1,825	2,020	2,154	2,257	41.5
Nonresident aliens as a percent of all students	13.38	17.85	19.43	19.82	19.77	[1]6.38
Other						
All students	43,512	38,617	42,341	44,225	46,305	6.4
U.S. students	40,282	34,667	37,696	39,021	40,819	1.3
Nonresident aliens	3,230	3,950	4,645	5,204	5,486	69.8
Nonresident aliens as a percent of all students	7.42	10.23	10.97	11.77	11.85	[1]4.42

[1] This figure represents the change in percentage points from 1980-81 to 1989-90.

NOTE—Data for 1980-81 and 1984-85 do not include imputations for nonresident aliens. Data for 1988-89 have been revised from previously published figures. Data represent programs, not organizational units within institutions. Because of adjustments to underreported and nonreported data, figures for degrees for all students for 1980-81, 1984-85 and 1986-87 in this table are slightly different from corresponding data in other tables. Data for Public affairs and services are different from data in the *Digest of Education Statistics* due to the use of different subcategories.

SOURCE: U.S. Department of Education, National Center for Education Statistics, HEGIS, "Degrees and Other Formal Awards Conferred" surveys; and IPEDS, "Completions" and "Consolidated" surveys.

Table 6-4.—Nonresident aliens receiving doctor's degrees, by field: 1980-81 to 1989-90

Field of study	1980-81	1984-85	1986-87	1988-89	1989-90	Percent change 1980-81 to 1989-90
All doctor's degrees						
All students	32,839	32,307	34,033	35,659	37,980	15.7
U.S. students	28,636	26,990	27,446	27,984	29,105	1.6
Nonresident aliens	4,203	5,317	6,587	7,675	8,875	111.2
Nonresident aliens as a percent of all students	12.80	16.46	19.35	21.52	23.37	[1]10.57
Agricultural sciences						
All students	1,067	1,216	1,048	1,183	1,272	19.2
U.S. students	724	892	717	741	772	6.6
Nonresident aliens	343	324	331	442	500	45.8
Nonresident aliens as a percent of all students	32.15	26.64	31.58	37.36	39.31	[1]7.16
Computer sciences						
All students	252	240	374	551	623	147.2
U.S. students	200	170	248	342	347	73.5
Nonresident aliens	52	70	126	209	276	430.8
Nonresident aliens as a percent of all students	20.63	29.17	33.69	37.93	44.30	[1]23.67
Engineering						
All students	2,551	3,174	3,818	4,523	4,965	94.6
U.S. students	1,595	1,776	2,036	2,331	2,474	55.1
Nonresident aliens	956	1,398	1,782	2,192	2,491	160.6
Nonresident aliens as a percent of all students	37.48	44.05	46.67	48.46	50.17	[1]12.70
Health sciences						
All students	842	1,172	1,213	1,436	1,543	83.3
U.S. students	754	1,033	1,025	1,210	1,276	69.2
Nonresident aliens	88	139	188	226	267	203.4
Nonresident aliens as a percent of all students	10.45	11.86	15.50	15.74	17.30	[1]6.85
Life sciences						
All students	3,718	3,354	3,417	3,520	3,844	3.4
U.S. students	3,429	2,978	2,890	2,956	3,092	-9.8
Nonresident aliens	289	376	527	564	752	160.2
Nonresident aliens as a percent of all students	7.77	11.21	15.42	16.02	19.56	[1]11.79
Physical sciences						
All students	3,140	3,382	3,671	3,858	4,168	32.7
U.S. students	2,610	2,700	2,696	2,721	2,860	9.6
Nonresident aliens	530	682	975	1,137	1,308	146.8
Nonresident aliens as a percent of all students	16.88	20.17	26.56	29.47	31.38	[1]14.50
Total - doctor's degrees in science						
All students	11,566	12,725	13,546	15,071	16,415	41.9
U.S. students	9,308	9,736	9,617	10,301	10,821	16.3
Nonresident aliens	2,258	2,989	3,929	4,770	5,594	147.7
Nonresident aliens as a percent of all students	19.52	23.49	29.00	31.65	34.08	[1]14.56
Mathematics						
All students	728	686	723	866	915	25.7
U.S. students	555	437	405	453	451	-18.7
Nonresident aliens	173	249	318	413	464	168.2
Nonresident aliens as a percent of all students	23.76	36.30	43.98	47.69	50.71	[1]26.95
Total - doctor's degrees in science and mathematics						
All students	12,294	13,411	14,269	15,937	17,330	41.0
U.S. students	9,863	10,173	10,022	10,754	11,272	14.3
Nonresident aliens	2,431	3,238	4,247	5,183	6,058	149.2
Nonresident aliens as a percent of all students	19.77	24.14	29.76	32.52	34.96	[1]15.18
Non-science and Mathematics Fields						
Business and management						
All students	844	849	1,094	1,149	1,142	35.3
U.S. students	683	646	779	837	751	10.0
Nonresident aliens	161	203	315	312	391	142.9
Nonresident aliens as a percent of all students	19.08	23.91	28.79	27.15	34.24	[1]15.16
Education						
All students	7,900	7,032	6,909	6,800	6,922	-12.4
U.S. students	7,307	6,434	6,323	6,228	6,312	-13.6
Nonresident aliens	593	598	586	572	610	2.9
Nonresident aliens as a percent of all students	7.51	8.50	8.48	8.41	8.81	[1]1.31

Table 6-4.—Continued

Field of study	1980-81	1984-85	1986-87	1988-89	1989-90	Percent change 1980-81 to 1989-90
Psychology						
All students	2,955	2,864	3,056	3,222	3,353	13.5
U.S. students	2,861	2,771	2,960	3,107	3,229	12.9
Nonresident aliens	94	93	96	115	124	31.9
Nonresident aliens as a percent of all students	3.18	3.25	3.14	3.57	3.70	[1]0.52
Public affairs and services						
All students	433	444	449	490	567	30.9
U.S. students	405	399	400	423	482	19.0
Nonresident aliens	28	45	49	67	85	203.6
Nonresident aliens as a percent of all students	6.47	10.14	10.91	13.67	14.99	[1]8.52
Social sciences						
All students	3,119	2,828	2,915	2,885	3,023	-3.1
U.S. students	2,701	2,231	2,295	2,159	2,268	-16.0
Nonresident aliens	418	597	620	726	755	80.6
Nonresident aliens as a percent of all students	13.40	21.11	21.27	25.16	24.98	[1]11.57
Other						
All students	5,294	4,879	5,341	5,176	5,643	6.6
U.S. students	4,816	4,336	4,667	4,476	4,791	-0.5
Nonresident aliens	478	543	674	700	852	78.2
Nonresident aliens as a percent of all students	9.03	11.13	12.62	13.52	15.10	[1]6.07

[1] This figure represents the change in percentage points from 1980-81 to 1989-90.

NOTE—Data for 1980-81 and 1984-85 do not include imputations for nonresident aliens. Data for 1988-89 have been revised from previously published figures. Data represent programs, not organizational units within institutions. Because of adjustments to underreported and nonreported data, figures for degrees for all students for 1980-81, 1984-85 and 1986-87 in this table are slightly different from corresponding data in other tables. Data for Public affairs and services are different from data in the *Digest of Education Statistics* due to the use of different subcategories.

SOURCE: U.S. Department of Education, National Center for Education Statistics, HEGIS, "Degrees and Other Formal Awards Conferred" surveys; and IPEDS, "Completions" and "Consolidated" surveys.

Table 6-5.— Postgraduation plans for nonresident aliens receiving doctor's degrees in the U.S.: 1980 to 1991

	Nonresident aliens receiving doctor's degrees			Nonresident aliens receiving doctor's degrees with definite plans in the U.S.								
					Postdoctoral study		Academic employment		Industrial employment		Other employment	
Year	Total	Percent with definite plans	Percent with definite plans in U.S.[1]	Total[1]	Number	Percent	Number	Percent	Number	Percent	Number	Percent
Science and engineering[2]												
1980	2,710	67.5	30.0	814	405	49.8	171	21.0	188	23.1	43	5.3
1985	4,028	62.6	30.5	1,230	610	49.6	345	28.0	223	18.1	45	3.7
1986	4,141	64.4	33.4	1,382	745	53.9	368	26.6	236	17.1	25	1.8
1987	4,450	64.2	33.6	1,494	871	58.3	332	22.2	230	15.4	51	3.4
1988	4,920	63.1	34.7	1,706	984	57.7	428	25.1	252	14.8	36	2.1
1989	5,378	63.5	35.8	1,928	1,108	57.5	453	23.5	302	15.7	54	2.8
1990	6,286	60.1	31.6	1,985	1,069	53.9	411	20.7	443	22.3	56	2.8
1991	7,252	58.5	35.2	2,551	1,341	52.6	536	21.0	570	22.3	92	3.6
Non-science and engineering[3]												
1980	934	66.3	9.9	92	24	26.1	56	60.9	7	7.6	5	5.4
1985	1,200	64.0	13.7	164	18	11.0	119	72.6	14	8.5	13	7.9
1986	1,135	66.9	18.9	214	35	16.4	152	71.0	12	5.6	14	6.5
1987	1,160	67.0	19.7	229	29	12.7	178	77.7	11	4.8	10	4.4
1988	1,275	64.4	19.8	253	40	15.8	196	77.5	10	4.0	5	2.0
1989	1,270	63.7	20.7	263	35	13.3	199	75.7	20	7.6	6	2.3
1990	1,458	59.7	22.8	332	42	12.7	258	77.7	22	6.6	9	2.7
1991	1,600	61.6	24.8	396	54	13.6	304	76.8	21	5.3	16	4.0
All fields												
1980	3,644	67.2	24.9	906	429	47.4	227	25.1	195	21.5	48	5.3
1985	5,228	62.9	26.7	1,394	628	45.1	464	33.3	237	17.0	58	4.2
1986	5,276	64.9	30.3	1,596	780	48.9	520	32.6	248	15.5	39	2.4
1987	5,610	64.8	30.7	1,723	900	52.2	510	29.6	241	14.0	61	3.5
1988	6,195	63.4	31.6	1,959	1,024	52.3	624	31.9	262	13.4	41	2.1
1989	6,648	63.5	33.0	2,191	1,143	52.2	652	29.8	322	14.7	60	2.7
1990	7,744	60.0	29.9	2,317	1,111	47.9	669	28.9	465	20.1	65	2.8
1991	8,852	59.0	33.3	2,947	1,395	47.3	840	28.5	591	20.1	108	3.7

[1] Includes nonresident aliens receiving doctor's degrees and staying in the U.S. with unknown plans of employment. These data are not shown separately.

[2] This category includes social sciences and psychology.

[3] This category includes unclassified.

Source: National Science Foundation, *Science and Engineering Doctorates: 1960-90*, pp. 219-224; and National *Science Foundation, Foreign Participation in U.S. Academic Science and Engineering: 1991*, p. 105.

Definitions

Agricultural sciences An instructional program that describes the principles and practices of agricultural research and production, and may prepare individuals to apply such knowledge and skills to the solution of practical agricultural problems. Includes instruction in basic animal, plant, and soil sciences; animal husbandry and plant cultivation; agribusiness and renewable natural resources, and soil conservation. Referred to in some source publications as "Agriculture and natural resources."

Area and ethnic studies A group of instructional programs that describe the history, society, politics, culture, and economics of a particular geographic region. Includes European studies, Middle Eastern studies, Scandinavian studies, etc.

Associate degree A degree granted for the successful completion of a sub-baccalaureate program of studies, usually requiring at least 2 years (or equivalent) of full-time college-level study. This includes degrees granted in a cooperative or work-study program.

Bachelor's degree A degree granted for the successful completion of a baccalaureate program of studies, usually requiring at least 4 years (or equivalent) of full-time college-level study. This includes degrees granted in a cooperative or work-study program.

Business and management Program of instruction that prepares individuals for a variety of activities in planning, organizing, directing, and controlling business office systems and procedures. Includes accounting, business administration, finance, and other related fields.

Carnegie unit A standard of measurement that represents one credit for the completion of a 1–year course.

College A postsecondary school which offers general or liberal arts education, usually leading to an associate, bachelor's, master's, doctor's, or first-professional degree. Junior colleges and community colleges are included under this terminology.

Computer sciences A group of instructional programs that describes computer and information sciences. Includes computer programming, data processing, information systems, and systems analysis. Referred to in some source publications as "Computer and information sciences."

Constant dollars Dollar amounts that have been adjusted by means of price and cost indexes to eliminate inflationary factors and allow direct comparison across years.

Consumer Price Index (CPI) This price index measures the average change in the cost of a fixed market basket of goods and services purchased by consumers.

Current dollars Dollar amounts that have not been adjusted to compensate for inflation.

Doctor's degree An earned degree carrying the title of Doctor. The Doctor of Philosophy degree (Ph.D.) is the highest academic degree and requires mastery within a field of knowledge and demonstrated ability to perform scholarly research. Other doctorates are awarded for fulfilling specialized requirements in professional fields, such as education (Ed.D.), musical arts (D.M.A.), business administration (D.B.A.), and engineering (D.Eng. or D.E.S.). Many doctor's degrees in academic and professional fields require an earned master's degree as a prerequisite. First-professional degrees, such as M.D. and D.D.S., are not included under this heading.

Earnings Includes wage and salary income, other labor income, and proprietor's income. Wage and salary income includes commissions, tips and bonuses. Other labor income includes employer contributions to private pension funds, welfare funds, and workmen's compensation insurance. Proprietor's income consists of not only monetary income, but also income-in-kind proprietorships and partnerships.

Education An instructional program that generally describes the theory and practice of learning and teaching. Includes teaching, educational administration, and special education.

Employment Includes civilian, noninstitutional persons who (1) worked during any part of the survey week as paid employees; worked in their own business, profession, or farm; or worked 15 hours or more as unpaid workers in a family-owned enterprise; or (2) were not working but had jobs or businesses from which they were temporarily absent due to illness, bad weather, vacation, labor-management dispute, or personal reasons whether or not they were seeking another job.

Engineering An instructional program that prepares individuals to apply mathematical and scientific principles to the solution of practical problems for the benefit of society. Includes aerospace, chemical, electrical, industrial, mechanical, nuclear, and petroleum engineering. Referred to in some source publications as "Engineering and related technologies."

Enrollment The total number of students registered in a given school unit at a given time, generally in the fall of a year.

Foreign languages A group of instructional programs that describes the structure and use of language that is common or indigenous to people of the same community or nation, the same geographical area, or the same cultural traditions. Programs cover such features as sound, literature, syntax, phonology, semantics, sentences, prose, and verse, as well as the development of skills and attitudes used in communicating and evaluating thoughts and feelings through oral and written language. Includes Asiatic, Germanic, Italic, and Semitic languages.

Full-time student A student enrolled in higher education courses with total credit load equal to at least 75 percent of the normal full-time course load.

Graduate An individual who has received formal recognition for the successful completion of a prescribed program of studies.

Graduate student The number of students who hold the bachelor's or first-professional degree, or the equivalent, and who are working towards a master's or doctor's degree. First-professional students are counted separately. These enrollment data measure those students who are registered at a particular time during the fall. At some institutions, graduate enrollment also includes students who are in postbaccalaureate classes but not in degree programs.

Health sciences An instructional program that prepares individuals to provide health care, or related research and support services, to individuals or groups. Referred to in some source publications as "Health professions." Includes such fields as audiology, dentistry, epidemiology, health services administration, medicine, nursing, optometry, pharmacy, public health, and veterinary medicine.

Higher education Study beyond secondary school at an institution that offers programs terminating in an associate, baccalaureate, or higher degree.

Higher education institutions

4–year institution An institution legally authorized to offer and offering at least a 4–year program of college-level studies wholly or principally creditable toward a baccalaureate degree.

2–year institution An institution legally authorized to offer and offering at least a 2–year program of college-level studies which terminates in an associate degree or is principally creditable toward a baccalaureate degree.

High school A secondary school offering the final years of high school work necessary for graduation, usually including grades 10, 11, 12 (in a 6–3–3 plan) or grades 9, 10, 11, and 12 (in a 6–2–4 plan).

High school program A program of studies designed to prepare students for their postsecondary education and occupation.

Humanities An instructional program that describes combined studies and research emphasizing humanistic subjects. Includes languages, literature, art, music, philosophy, and religion.

Labor force (or Workforce) Persons employed as civilians, unemployed (but looking for work), or in the armed services during the survey week. The "civilian labor force" comprises all civilians classified as employed or unemployed.

Law (Legal Studies) A summary of groups of instructional programs that describe the theory, history, and application of the rules of conduct by which societal relations are formally structured and adjudicated. (In this publication, the term refers to legal studies, as law degrees are first professional degrees which are not covered in this publication.)

Liberal arts and sciences An instructional program that describes a structured combination of the arts, biological and physical sciences, social sciences, and humanities, emphasizing breadth of study.

Life sciences An instructional program that describes the scientific study of living organisms and their systems. Referred to in some source publications as "biological sciences." Includes biology, botany, microbiology, and zoology.

Master's degree A degree awarded for successful completion of a program generally requiring 1 or 2 years of full-time college-level study beyond the bachelor's degree. One type of master's degree, including the Master of Arts degree, or M.A., and the Master of Science degree, or M.S., is awarded in the liberal arts and sciences for advanced scholarship in a subject field or discipline and demonstrated ability to perform scholarly research. A second type of master's degree is awarded for the completion of a professionally oriented program, for example, an M.Ed. in education, an M.B.A. in business administration, an M.F.A. in fine arts, an M.M. in music, an M.S.W. in social work, and an M.P.A. in public administration. A third type of master's degree is awarded in professional fields for study beyond the first-professional degree, for example, the Master of Laws (L.L.M.) and Master of Science in various medical specializations.

Mathematics An instructional program that describes the rigorous analysis of quantities, magnitudes, forms, and their relationships, using symbolic logic and language. Includes instruction in actuarial sciences, algebra, calculus, functional analysis, geometry, number theory, logic, statistics, topology, and other mathematical specializations.

Nonresident alien A person who is not a citizen of the United States and who is in this country on a temporary basis and does not have the right to remain indefinitely.

Part-time student A student enrolled in higher education courses with a total credit load less than 75 percent of the normal full-time credit load.

Philosophy and religion A summary of groups of instructional programs that describe the study of modes, methods and types of logical inquiry; and the study of organized systems of belief and related practices.

Physical sciences An instructional program that describes the scientific study of inanimate objects, processes of matter and energy, and associated phenomena. Includes astronomy, astrophysics, atmospheric sciences, chemistry, geology, oceanography, physics, and science technologies.

Protective services A summary of groups of instructional programs that describe the principles and procedures for providing police, fire, and other safety services, and for managing penal institutions. Includes criminology and fire protection.

Psychology A summary of groups of instructional programs that describe the scientific study of the behavior of individuals, independently or collectively, and the physical and environmental bases of mental, emotional, and neurological activity. Includes clinical psychology, counseling, educational psychology, and industrial psychology.

Public affairs A group of instructional programs that prepare individuals to analyze, manage, and deliver public programs and services. This group includes public administration, public policy analysis, social work, and other related areas. Referred to in some tables as "Public affairs and services."

Region The geographic regions used in this publication are as follows:

Northeast - Connecticut, Maine, Massachusetts, New Hampshire, New Jersey, New York, Pennsylvania, Rhode Island, Vermont.

Midwest - Illinois, Indiana, Iowa, Kansas, Michigan, Minnesota, Missouri, Nebraska, North Dakota, Ohio, South Dakota, Wisconsin.

South - Alabama, Arkansas, Delaware, District of Columbia, Florida, Georgia, Kentucky, Louisiana, Maryland, Mississippi, North Carolina, Oklahoma, South Carolina, Tennessee, Texas, Virginia, West Virginia.

West - Alaska, Arizona, California, Colorado, Hawaii, Idaho, Montana, Nevada, New Mexico, Oregon, Utah, Washington, Wyoming.

Salary The total amount regularly paid or stipulated to be paid to an individual, before deductions, for personal services rendered while on the payroll of a business or organization.

Science The body of related courses concerned with knowledge of the physical and biological world and with the processes of discovering and validating this knowledge. In this publication, unless otherwise noted, referring to aggregate data for the following fields: agricultural sciences, computer sciences, engineering, health sciences, life sciences and physical sciences.

Social sciences An instructional program that describes the study of human social behavior and social institutions. These studies include anthropology, archaeology, criminology, economics, geography, history, political science, sociology, and urban studies.

Student An individual for whom instruction is provided in an educational program under the jurisdiction of a school, school system, or other education institution. No distinction is made between the terms "student" and "pupil," though "student" may refer to one receiving instruction at any level while "pupil" refers only to one attending school at the elementary or secondary level. A student may receive instruction in a school facility or in another location, such as at home or in a hospital. Instruction may be provided by direct student-teacher interaction or by some other approved medium such as television, radio, telephone, and correspondence.

Undergraduate students Students registered at an institution of higher education who are working in a program leading to a baccalaureate degree or other formal award below the baccalaureate, such as an associate degree.

Underemployed In this publication this word is used to describe recent college graduates who indicated that a college degree was not required for their job and who were employed full-time in sales, service, administrative support/clerical, or craft/operator/laborer positions.

Unemployed Individuals who were not working for pay, were looking for work, and were available to work.

U.S. Service Schools These institutions of higher education are controlled by the U.S. Department of Defense and the U.S. Department of Transportation. The ten institutions counted in the NCES surveys of higher education institutions include: the Air Force Institute of Technology, Community College of the Air Force, Naval Postgraduate School, Uniformed Services University of the Health Sciences, U.S. Air Force Academy, U.S Army Command and General Staff College, U.S. Coast Guard Academy, U.S. Merchant Marine Academy, U.S. Military Academy, and the U.S. Naval Academy.

Guide to Sources

Sources and Comparability of Data

The information presented in this report was obtained from the National Center for Education Statistics (NCES), the National Science Foundation (NSF), the Bureau of the Census, and the Bureau of Labor Statistics. Users should take particular care when comparing data from different sources. Differences in procedures, timing, phrasing of questions, interviewer training, and so forth mean that the results from the different sources may not be strictly comparable. Following the general discussion of data accuracy below, descriptions of the information sources and data collection methods are presented, grouped by sponsoring organization. More extensive documentation of a particular survey's procedures does not imply more problems with the data, only that more information is available.

Accuracy of Data

The accuracy of any statistic is determined by the joint effects of "sampling" and "nonsampling" errors. Estimates based on a sample will differ somewhat from the figures that would have been obtained if a complete census had been taken using the same survey instruments, instructions, and procedures. In addition to such sampling errors, all surveys, both universe and sample, are subject to design, reporting, and processing errors and errors due to nonresponse. To the extent possible, these nonsampling errors are kept to a minimum by methods built into the survey procedures. In general, the effects of nonsampling errors are more difficult to gauge than those produced by sampling variability.

National Center for Education Statistics (NCES) Surveys

Integrated Postsecondary Education Data System

The Integrated Postsecondary Education Data System (IPEDS) surveys all postsecondary institutions, including universities and colleges, as well as institutions offering technical and vocational education beyond the high school level. This survey, which began in 1986, replaces and supplements the Higher Education General Information Survey (HEGIS).

The IPEDS consists of several integrated components that obtain information on who provides postsecondary education (institutions), who participates in it and completes it (students), what programs are offered and what programs are completed, and both the human and financial resources involved in the provision of institutionally-based postsecondary education. Specifically, these components include: institutional characteristics, including institutional activity; fall enrollment, including age and residence; fall enrollment in occupationally-specific programs; completions; finance; staff; salaries of full-time instructional faculty; and academic libraries.

Prior to the establishment of IPEDS in 1986, HEGIS acquired and maintained statistical data on the characteristics and operations of institutions of higher education. Implemented in 1966, HEGIS was an annual universe survey of institutions listed in the latest NCES *Education Directory, Colleges and Universities.*

The tables presented in this report draw on HEGIS surveys which solicited information concerning institutional characteristics, faculty salaries, finances, enrollment, and degrees. Since these surveys were distributed to all higher education institutions, the data presented were not subject to sampling error. However, they were subject to nonsampling error, the sources of which varied with the survey instrument. Information concerning the nonsampling error of the enrollment and degrees surveys draws extensively on the "HEGIS Post-Survey Validation Study" conducted in 1979. In some cases minor revisions have been made to reflect tabulation procedures more consistent with current definitions.

In the IPEDS system, institutions are counted in the state of their physical location, in instances where institutions have multi-state campuses. In the former HEGIS system, institutions were counted in the state where the main campus of an institution was located. For this publication, the HEGIS data have been adjusted to conform to the newer IPEDS state location definitions. Therefore, the data for some states may differ from data in the *Digest of Education Statistics* and other NCES publications.

Institutional Characteristics

This survey provided the basis for the universe of institutions presented in the *Education Directory, Colleges and Universities*. The universe comprised institutions that met certain accreditation criteria and offered at least a 1–year program of college-level studies leading toward a degree. All of these institutions were certified as eligible by the U.S. Department of Education's Division of Eligibility and Agency Evaluation. Each fall, institutions listed in the previous year's *Directory* were asked to update a computer printout of their information.

Fall Enrollment

This survey has been part of the IPEDS or HEGIS series since 1966. The enrollment survey response rate has been relatively high; the 1989 response rate was 86.1 percent. Major sources of nonsampling error for this survey were classification problems, the unavailability of needed data, interpretation of definitions, the survey due date, and operational errors. Of these, the classification of students appears to have been the main source of error. Institutions had problems in correctly classifying first-time freshmen, other first-time students, and unclassified students for both full-time and part-time categories. These problems occurred most often at 2–year institutions (private and public) and private 4–year institutions. In the 1977–78 HEGIS validation studies, the classification problem led to an estimated overcount of 11,000 full-time students and an undercount of 19,000 part-time students. Although the ratio of error to the grand total was quite small (less than 1 percent), the percentage of errors was as high as 5 percent for detailed student levels and even higher at certain aggregation levels.

Beginning with fall 1986, the survey system was redesigned with the introduction of IPEDS (see above). The new survey system comprises all postsecondary institutions, but also maintains comparability with earlier surveys by allowing HEGIS institutions to be tabulated separately. The new system also provides for preliminary and revised data releases. This allows the Center flexibility to release early data sets while still maintaining a more accurate final data base.

Completions

This survey was part of the HEGIS series throughout its existence. However, the degree classification taxonomy was revised in 1970–71 and 1982–83. Collection of degree data has been maintained through the IPEDS system. The nonresponse rate did not appear to be a significant source of nonsampling error for this survey. The response rate for all higher education and noncollegiate institutions for the 1988–89 survey was 76.3 percent. Because of the high return rate for higher education data used in this report, nonsampling error caused by imputation was minimal.

The major sources of nonsampling error for this survey were differences between the NCES program taxonomy and taxonomies used by the colleges, classification of double majors and double degrees, operational problems, and survey timing. In the 1979 HEGIS validation study, these sources of nonsampling error were found to contribute to an error rate of 0.3 percent overreporting of bachelor's degrees and 1.3 percent overreporting of master's degrees.

Questions concerning the surveys used as data sources for this report or other questions concerning HEGIS and IPEDS can be directed to:

Postsecondary Education Statistics Division
National Center for Education Statistics
555 New Jersey Avenue NW
Washington, DC 20208–5652

The 1990 High School Transcript Study

The 1990 High School Transcript Study was conducted by Westat, Inc. for NCES. Approximately 21,500 transcripts were collected from 330 schools. Tabulations from this transcript study, the 1987 study and the 1982 study were used in the publication: *Comparative Data on Credits Earned and Demographics for 1990, 1987, and 1982 High School Graduates.*

The 1987 data are based on approximately 22,700 transcripts obtained as part of the 1987 High School Transcript Study. The 1982 data are based on approximately 12,000 transcripts collected by the High School and Beyond Study. All three studies coded the courses taken by students using the Classification of Secondary School Courses (CSSC).

Further information on this survey may be obtained from:

Patricia Dabbs
Education Assessment Division
National Center for Education Statistics
555 New Jersey Avenue NW
Washington, DC 20208–5652

Survey of Recent College Graduates

NCES has conducted periodic surveys of persons, about 1 year after graduation, to collect information on college outcomes. The "Recent College Graduates" surveys have concentrated on those graduates entering the teaching profession. To obtain accurate results on this subgroup, graduates who are newly qualified to teach have been oversampled in each of the surveys. The survey involves a two-stage sampling procedure. First, a sample of institutions awarding bachelor's and master's degrees is selected and stratified by percentage of education graduates, control, and type of institution. Second, for each of the selected institutions, a sample of degree recipients is chosen. The response rates on the Recent College Graduates survey have tended to be low because of the great difficulty in tracing the students after graduation. Much more of the nonresponse can be attributed to invalid mailing addresses than to refusals to participate. Despite their shortcomings, the data are presented in this report because they provide valuable information not available elsewhere about college outcomes. Users should be cautious about drawing conclusions based on data from small samples. It is also likely that the data are somewhat biased since the more mobile students, such as graduate students, are the most difficult to track for the survey.

The 1985 survey requested data from 18,738 graduates from 404 colleges. Responses were obtained from 13,200 students, for a response rate of 74 percent (885 were out of scope). The response rate for the colleges was 98 percent. The 1987 survey form was sent to 21,957 graduates. Responses were received from 16,878, for a response rate of 79.7 percent.

The 1991 RCG study involved a sample of 18,000 graduates of 400 bachelor's and master's degree-granting institutions. The 18,000 graduates consisted of 16,000 bachelor's degrees recipients and 2,000 master's degree recipients between July 1, 1988 and June 30, 1989. The response rate was 83.2 percent, higher than the last survey. Random samples of graduates were selected from each field of study. Graduates in education, mathematics, and the physical sciences were sampled at a higher rate, as were minority graduates to provide a sufficient number of these graduates for analysis purposes. The graduates included in the sample were selected in proportion to the institution's number of graduates. The samples are drawn from the universe of students within 1 year of attaining a bachelor's or master's degree.

Further information on this survey may be obtained from:

Peter Stowe
Postsecondary Education Statistics Division
National Center for Education Statistics
555 New Jersey Avenue NW
Washington, DC 20208–5652

Publications

Comparative Data on Credits Earned and Demographics for 1990, 1987, and 1982 High School Graduates. See The 1990 High School Transcript Study above.

The Condition of Education

The Condition of Education is an annual indicators publication which contains data on a limited number of indicators describing major topics of interest and concern in education today. No more than 60 indicators are presented in each year's report. These indicators represent a consensus of professional judgement on the most significant national measures of the condition and progress of education at a given time. The indicators include a basic core that can be repeated with information every year, supplemented by a more limited set of indicators based on infrequent or one-time studies. Academic subject categories used for the *Condition* data that are presented in *Degrees in Science and Mathematics: National Trends and State-By-State Data* are described on pages 250 through 265 of the 1992 edition of the *Condition.*

For more information on *The Condition of Education* contact:

Nabeel A. Alsalam
Data Development Division
National Center for Education Statistics
555 New Jersey Avenue, NW
Washington, DC 20208–5650

Digest of Education Statistics

The *Digest of Education Statistics* is the primary resource publication on education statistics published by NCES. Its primary purpose is to provide a compilation of statistical information covering the broad field of American education from kindergarten through graduate school. To qualify for inclusion in this publication, material must be nationwide in scope and of current interest and value. The *Digest* is now divided into seven chapters: All Levels of Education, Elementary and Secondary Education, Postsecondary Education, Federal Programs for Education and Related Activities, Outcomes of Education, International Comparisons of Education, and Learning and Technology. Each chapter contains an introduction to statistical materials describing the most significant data in the chapter. Charts are provided to further illuminate important data. The *Digest* includes a selection of data from many sources, both government and private, and draws especially on the results of surveys and activities carried out by NCES. Academic subject categories used for the *Digest* data that are presented in *Degrees in Science and Mathematics: National Trends and State-By-State Data* are described on pages 243 through 261 of the 1992 edition of the *Digest.*

For more information on the *Digest of Education Statistics* contact:

Thomas D. Snyder
Data Development Division
National Center for Education Statistics
555 New Jersey Avenue, NW
Washington, DC 20208–5650

Occupational and Educational Outcomes of Recent College Graduates 1 Year After Graduation: 1991 and Occupational and Educational Outcomes of 1985–86 Bachelor's Degree Recipients 1 Year After Graduation: 1987

Reports presenting results of the *Survey of Recent College Graduates* survey, (RCG). See *Survey of Recent College Graduates* above.

Race/Ethnicity Trends in Degrees Conferred by Institutions of Higher Education: 1980–81 through 1989–90

This report presented data on degrees conferred compiled from IPEDS (see above), in racial/ethnic categories, by level of degree and field. Academic subject categories used for the data in this *Race/Ethnicity Trends in Degrees Conferred by Institutions of Higher Education: 1980–81 through 1989–90* that are presented in *Degrees in Science and Mathematics: National Trends and State-By-State Data* are described in this publication on pages 31 - 46.

Questions concerning this report can be directed to:

Frank B. Morgan
Postsecondary Education Statistics Division
National Center for Education Statistics
555 New Jersey Avenue NW
Washington, DC 20208–5652

Who Majors in Science? College Graduates in Science, Engineering, or Mathematics from the High School Class of 1980.

This report examined the major fields of study of a representative sample of 1980 high school seniors who had graduated from college by 1986 and compared those who majored in engineering, mathematics, or the natural and physical sciences with those who majored in other fields.

Questions concerning this report can be directed to:

Thomas Snyder
National Center for Education Statistics
555 New Jersey Avenue NW
Washington, DC 20208–5652

National Science Foundation Publications

Science and Engineering Indicators

Science and Engineering Indicators is a biennial publication of the National Science Board, a part of the National Science Foundation. This report informs researchers and policy makers in both the public and private sectors on the status of U.S. science and engineering education and research. *Science and Engineering Indicators* provides data about the U.S. technology base. It consists of seven chapters: Precollege Science and Mathematics Education, Higher Education in Science and Engineering, Science and Engineering Workforce, Financial Resources for Research and Development, Academic Research and Development: Financial Resources, Personnel, and Outputs, Technology and Global Competitiveness, Attitudes Toward Science and Technology: The United States and International Comparisons. Academic subject categories used for the data in *Science and Engineering Indicators* presented in *Degrees in Science and Mathematics: National Trends and State-By-State Data* can be found in the Appendix.

For more information contact:
National Science Foundation
1800 G Street NW
Washington, DC 20550

Science and Engineering Doctorates: 1960–90

Science and Engineering Doctorates: 1960–90 is a report presenting data from the Survey of Earned Doctorates (SED). The SED has collected basic statistics from the universe of doctoral recipients in the United States each year since the 1920s. It has been supported by five Federal agencies: the National Science Foundation, in conjunction with the U.S. Department of Education; the National Endowment for the Humanities; the U. S. Department of Agriculture; and the National Institutes of Health.

A survey form is distributed, with the assistance of graduate deans, to each person completing the requirements for a doctorate. Of the approximately 31,000 persons eligible for the survey, approximately 95 percent respond. The questionnaire obtains information on sex, race/ethnicity, marital status, citizenship, handicaps, dependents, specialty field of doctorate, educational institutions attended, time spent in completion of doctorate, financial support, educational debt, postgraduation plans, and educational attainment of parents. The data are collected, edited, and published by the National Academy of Sciences.

For further information contact:
Susan Hill
National Science Foundation
1800 G Street NW
Room L609
Washington, DC 20550

Foreign Participation in U.S. Academic Science and Engineering: 1991

This report presents data on the distribution of foreign students in American institutions of higher education, focusing on students in science and engineering. Information includes changes in numbers of foreign students in American graduate scientific and engineering programs, and data on foreign members of the American scientific and engineering community. The data obtained from this report on postgraduate plans on nonresident aliens who were awarded doctor's degrees from U.S. institutions were from the Doctorate Records File of the NSF's Division of Science Resources Studies.

For further information contact:
Susan Hill
National Science Foundation
1800 G Street NW
Room L609
Washington, DC 20550

Bureau of the Census

Current Population Survey

The monthly Current Population Survey (CPS) sample consists of 729 areas comprising 1,973 counties, independent cities, and minor civil divisions throughout the 50 States and the District of Columbia. The sample was initially selected from the 1980 census files and is periodically updated to reflect new housing construction.

The monthly CPS deals primarily with labor force data for the civilian noninstitutional population (i.e., excluding military personnel and their families living on post and inmates of institutions). The estimation procedure employed for the monthly CPS data involves inflating weighted sample results to independent estimates of characteristics of the civilian noninstitutional population in the United States by age, sex, and race. These independent estimates are based on statistics from decennial censuses; statistics on births, deaths, immigration, and emigration; and statistics on the population in the armed services. Generalized standard error tables are provided in the *Current Population Reports*. The data are subject to both nonsampling and sampling errors.

Further information is available in the *Current Population Reports*. Series P-20, or by contacting:
Education and Social Stratification Branch
Population Division
Bureau of the Census
U.S. Department of Commerce
Washington, DC 20233

Bureau of Labor Statistics

Occupational Outlook Handbook

The *Occupational Outlook Handbook* is published on a biennial basis by the Bureau of Labor Statistics. This publication presents projections of the labor force, economic growth, industry output, and employment under various assumptions. The report identifies the factors affecting job prospects and describes the interaction of these factors with the future growth of the occupation. The *Occupational Outlook Handbook* presents 250 statements on specific occupations, presenting current statistics and expected job growth through the year 2005.

For more information contact:
Bureau of Labor Statistics
U.S. Department of Labor
Washington, DC 20212

ISBN 0-16-042071-7